Patterns of change in tropical plants

To G H C and A F C

'Botanical teaching based on the temperate flora must necessarily be ill-balanced and inadequate.'

C G G J Van Steenis (1962)

Patterns of change in tropical plants

G P Chapman, B.SC. PH.D.

University of London Press Ltd

ISBN 0 340 11932 2

University of London Press Ltd
St Paul's House, Warwick Lane, London EC4

Printed and bound by
Richard Clay (The Chaucer Press), Ltd, Bungay, Suffolk

Contents

Preface

The work of Darwin and others during the nineteenth century combined with that following from the rediscovery of Mendel's papers during the early twentieth century has led to an impressive understanding of evolutionary processes. Many of the conclusions about plant evolution are based largely though not exclusively on temperate groups, even though the flora of the tropics is both larger and more varied. Such a situation is certainly stimulating but at the same time unsatisfactory in that it creates difficulties for those who make their first serious acquaintance with botany in the tropics. While not wishing to exclude a wider audience, I considered my first obligation in writing this book to be towards such students.

My aims have been to show something of the value of cytogenetics in understanding the processes of plant speciation and in the case of crop plants to indicate where these processes can be turned to our advantage. I have assumed on the part of the reader some knowledge of genetics and cytology and the material here is complementary to other works dealing with experimental taxonomy, plant breeding and evolution.

Opposite the title page appears a comment of Professor Van Steenis that is germane to botanical teaching in the tropics even more than it is in other regions. As may be appreciated, to present some kind of logical and balanced treatment in a short book involves one in all manner of problems of inclusion and exclusion. The result is one author's bias and for that the surest remedy is more authors who, variously irked or encouraged by what they read, produce their own corrective.

The word 'tropical' has been liberally interpreted here, and examples from the sub-tropics and even further afield are included. I would justify this partly by the relative paucity of examples that have been thoroughly worked out for tropical latitudes. An instructive problem of selection is provided by the genus *Casuarina* whose natural distribution runs through Malaysia and Australia and for which most data is available in the latter area. By drawing attention to *C. nana* perhaps I can encourage interest in the less well-known tropical species.

While writing this book I have had all manner of help from colleagues which is acknowledged elsewhere. The subject of tropical plant evolution is a large one. This book seeks only to introduce it.

G P Chapman
University of the West Indies
Mona, Jamaica
December 1968

Foreword

Since 1945 scientists have shown a growing awareness that an under-standing of tropical biology is essential before we can comprehend the general principles of evolution and ecology in the world as a whole. The historical accident that, until recently, the centres of research and publication of biological works have all been located in temperate regions has led to the propagation of 'laws' and 'rules' that are based on the specialised vegetation that survives in the stressful temperate re-gions. Often these are applicable only with difficulty to the flora of the tropics. How much more logical it would have been if we had begun our biology in the tropics and then refined it to take account of the special problems posed by the diurnal and seasonal fluctuations in tem-perature and photo-period of higher latitudes and the more complicated climatic history outside of the tropics. How much wiser we should have been if we had foreseen that the most difficult agricultural problems facing a burgeoning human population would be those of tropical lands where technological methods that are manifestly successful in temperate countries have little relevance.

Only belatedly are biologists awakening to face the truth—that there is an enormous amount of work to be done before we can appreciate what evolution has wrought in the tropics and how it has come about. They are also discovering that the necessary research is fascinating to carry out and highly rewarding in the satisfaction that comes from tack-ling a problem that was crying out for solution. Their reward is much greater than if they were competing feverishly with a multitude of colleagues in the hope of discovering today what a half dozen others will discover tomorrow in the same temperate taxon. Geoffrey Chapman is one of the young men who has accepted the challenge of tropical biology, has been successful in the field and is anxious to make it possible for others to join him. *Patterns of Change in Tropical Plants* should do just this.

Herbert G Baker
University of California
Berkeley
23 May 1970

Acknowledgments

My thanks are due to Professor A D Skelding, Dr N W Simmonds, Mr D Ingle Smith, Dr J B Free, Professor J R Harlan, Dr P Hunt and Mr K Shepherd, who have read critically various parts of the book, and to the various librarians who have helped minimise the effects of living on a tropical island. My thanks are also due to Dr S Sehgal, who not only provided specialist help and data on maize but gave helpful advice and constructive criticism throughout, to Professor H G Baker, who contributed the Foreword, and to Dr C D Adams both for advice and for data from his forthcoming *Flora of Jamaica*. It is also a pleasure to acknowledge help from Professor J G Hawkes, who read the book in proof and gave advice and comment.

For permission to use various illustrations and extracts from other works I have to thank Messrs R Speak, A H C Carter and Longmans Green Ltd, Oliver and Boyd Ltd, Professor C D Darlington, the American Genetic Association and the Journal *Heredity*, the Society for the Study of Evolution and the *Journal of Evolution*, Professor J R Harlan, the University of Chicago Press and the *Botanical Gazette*, Professor P Mangelsdorf and the editors of *Science*, copyright 13.3.69 by the American Association for the Advancement of Science, the Smithsonian Institute, Professor J Heslop Harrison and Heinemann Educational Books Ltd, the Bombay Natural History Society, Methuen and Company Ltd, Dr W Brown of the Pioneer Hi-bred Seed Company and Constance Dear for the illustrations on pp 17, 19, 36, 45, 59 and 77. I am also indebted for much editorial help from Mrs J Danaher.

I also thank my wife, whose typing ensured that what was written might also be read, for her encouragement.

Chapter 1

The tropical environment

The impression made upon a biologist visiting the tropics for the first time must, almost invariably, be one of surprise at the extraordinary numbers of species and at the wide range of plant and animal habitats which are to be found. Even small islands, provided they have a varied topography, accommodate a wealth of species which seems disproportionate to their size. There are, as might be expected, large areas of apparently uniform vegetation such as the tropical rain forest of the Amazon basin but, compared with temperate regions, these have a much wider representation of tree species which in turn create more habitats for smaller plants.

Tropical countries are in many instances mountainous and the effects of lower temperature and higher rainfall are evident. Where mountains rise to 4000 metres and beyond, it is possible to find most of the environmental conditions (apart from extreme day-length variation) which exist at higher latitudes and a given region may contain, in addition to more obviously tropical plants, a montane flora of distinctly temperate character. To define tropical plants as those growing between latitudes 23° 27′ North (Cancer) and 23° 27′ South (Capricorn) ignores not only the effects of altitude but those too, of ocean currents which, if they are moving away from the equator, will warm the various land masses that lie across their path. Because of the Gulf Stream, the Bahamas, for instance, have a warmer climate than, say, the island of San Ambrosio lying at a comparable southern latitude off the coast of Chile but cooled by the Humboldt current. Even more surprising is the occurrence of what is virtually 'tropical' rain forest beyond the tropics in eastern Australia, north Burma and eastern Brazil (Richards, 1952). The purpose here, however, is not to produce the perfect definition of 'tropical' but rather to raise questions about vegetation.

Evolution and the tropics

Plant evolution in the tropics has produced all manner of striking ɛdaptation. It has become traditional for writers of botany textbooks to emphasise particular points by reference to extreme examples which turn out almost invariably to be tropical. The impression is thereby fostered that tropical plants are somehow odd. It is probably at least as realistic to regard temperate vegetation, with its narrower range of species and its adaptation to prolonged dormancy, as the comparatively threadbare fringe of the world's flora.

The theory of evolution gained impetus when Darwin visited the tropics and saw the effects of isolation at first hand. Isolation is a potent factor in the development of new varieties and species and often, though not invariably, this is associated with islands. In 1835 Darwin reached the Galapagos, a group of tropical islands west of Ecuador, and was impressed by the differences among the animals and plants from the different islands. In *The Voyage of the Beagle* (1839) he wrote:

'I have not as yet noticed by far the most remarkable feature in the natural history of this archipelago; it is, that the different islands to a considerable extent are inhabited by a different set of beings. My attention was first called to this fact by the Vice-Governor, Mr Lawson, declaring that the tortoises differed from the different islands, and that he could with certainty tell from which island any one was brought.'

He returned to this idea in the *Origin of Species* (1859) and there is no doubt that what he saw in the Galapagos significantly influenced the development of his theory of evolution.

Tropical vegetation types

On the larger land masses there is a gradation from equatorial rain forest to arid desert or high alpine conditions and to define 'types' of vegetation is to some extent an arbitrary procedure. On a world basis Good (1964) classified plants into six kingdoms and thirty-seven regions. A simpler approach is that of Polunin (1960) who, within the tropics, recognises the major types upon which the following summary is largely based.

Tropical rain forest

This is the most luxuriant of all vegetation and is confined to lowland tropical or near-tropical regions with abundant rainfall. Temperatures are high and uniform at 25–26° C, rainfall 200–400 cm annually and with about 80 per cent humidity. The plant complex consists of tall trees more or less referable to three layers, the tallest reaching about thirty metres. Around these trees are herbs, climbers, lianes, epiphytes, stranglers, saprophytes and parasites, and flowering, fruiting and leaf replacement tend to be continuous rather than seasonal processes. Where there is a more or less marked seasonality the luxuriance of the vegetation is checked and under conditions of decreasing moisture gives place to monsoon forest, savanna woodland or thorn woodland.

Monsoon forest

Where rainfall amounts to between 100 and 200 cm per annum with a pronounced dry period of four to six months the vegetation type is that of 'monsoon forest'. During the dry season the trees may defoliate, and throughout the year the daily variation of temperature is more marked than occurs in tropical rain forest.

Trees are more widely spaced and there is no conspicuous division into three-layered canopy. Beneath the canopy there are, typically, shrub and ground layers. The epiphyte and other tree-associated species are much less conspicuous than in tropical rain forest.

Savanna woodland

Under conditions somewhat drier than those of monsoon forest the vegetation is again more open with widely spaced trees, except perhaps locally along the watercourses. Lianes and epiphytes have only scant representation, but grasses and xerophilous shrubs are more conspicuous elements in the vegetation.

Thorn woodland

Such an environment will develop where the rainfall is between 40 and 80 cm annually. Foliage tends to be much reduced and xerophilous so that plants with thorns, prickles and scales are common. Grasses are a relatively minor part of the vegetation, and many plants have water-conserving devices. If such rain as does fall is strongly seasonal this will be reflected in short periods of intensive growth and flowering.

Desert

Where rainfall is minimal, plant life may be reduced to vanishing point. Those species which do survive are adapted in an extreme degree to water conservation. Certain ephemeral species avoid drought by confining the active part of their life cycle to the brief periods of rainfall.

The habitats which have been described merge into each other, but further modifications are imposed by the effects of altitude. With increased elevation temperatures are lower and the humidity is usually, though not invariably, higher. Richards (ibid) distinguishes for the mountain regions of Malaysia a sub-montane rain forest (500–1000 m), montane rain forest including elfin woodland (1000–2400 m) and tropical sub-alpine forest (2400–4000 m), and these types of vegetation can be found in other corresponding types of habitat. Lastly, in addition to those habitats already mentioned, there are the specialised ones such as mangrove swamps and the littoral communities which develop between and near the tidal extremes.

Within the foregoing vegetation types the cytogenetic changes take place which form the subject of this book. Before proceeding to discuss them, however, it is necessary to examine briefly some of the more potent factors which influence or have influenced the plant life of these regions.

Seasonal change

Seasonal change is not very conspicuous in the tropical rain forest but becomes more evident in monsoon forest. Since these and other vegetation types can occur at the same latitude, the differences are partly explainable in terms of aspect or continentality. Such differences, however, only accentuate the consequences ultimately attributable to the

TABLE 1. DURATION OF DAY LENGTH IN VARIOUS LATITUDES AT DIFFERENT TIMES OF THE YEAR

Latitude	21 March		21 June		21 September		21 December	
	Hr	Min.	Hr	Min.	Hr	Min.	Hr	Min.
60° N	12	18	18	53	12	23	5	52
30° N	12	09	14	05	12	10	10	12
0°	12	07	12	07	12	06	12	07

Note: Especially at high latitudes refractive effects of the earth's atmosphere prolong day length for several minutes.

Data by permission of Smithsonian Institution

earth's rotations. The earth rotates on its axis so that all parts of its surface are alternately turned to and from the sun, thereby causing day and night. Because the axis of this rotation is tipped, places at different distances from the equator need not have the same day length (see table 1). It is evident that at zero latitude day length is virtually constant throughout the year and that the March and September dates show about twelve hours at all latitudes. It is also true that, while the hours of daylight are distributed differently in various latitudes, their annual totals are comparable.

Figure 1 Diagram showing the positions of the earth relative to the sun at various times during its annual rotation. Seasons are shown for the northern hemisphere.

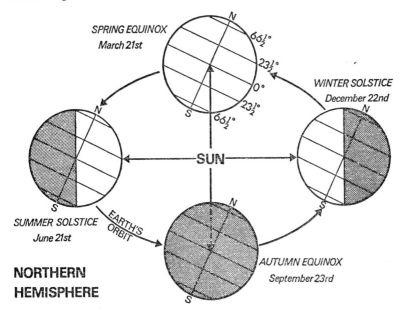

The inclined axis of the earth means that at different times in its annual rotation the poles are turned alternately towards the sun (see figure 1). In the northern hemisphere when the pole is turned away from the sun from September to March longer periods of darkness and diminished heat supply result. The direction of the earth's axis relative to the sun has comparatively little effect in the tropics, partly because a constant day length ensures a more or less consistent heat supply and partly for other reasons which will now be examined.

People living north or south of the tropics see the sun rising higher in the sky each midday to a maximum at the summer solstice (22 June in the north and 22 December in the south). Within the tropics, however, the sun is seen to reach its maximum height in the sky (actually overhead or 90°) *twice* during the year. Table 2 shows the dates at which the sun is overhead at noon for various latitudes.

TABLE 2. DATES OF OVERHEAD SUN FOR VARIOUS LATITUDES

Latitude	Date		
23° 27′ (Cancer)	22 June		
20°	21 May	24 July	
10°		16 Apr	28 Aug
0°		21 Mar	23 Sept
−10°		23 Feb	20 Oct
−20°	21 Jan		22 Nov
−23° 27′ (Capricorn)			22 Dec

Note: Dates vary slightly from year to year, the figures quoted referring to 1950.

Data by permission of Smithsonian Institution

In the tropics the sun is seen in a more nearly vertical position relative to the observer than at higher latitudes with the important result that the angle of incidence of the sun's rays is steeper (see figure 2). For Cambridge (England) at latitude 52° N the angle of the sun's rays varies between $61\frac{1}{2}$° on 22 June and $14\frac{1}{2}$° on 22 December. At the equator the sun's angle varies from 90° on 21 March and 23 September to 66° on 22 June and 22 December. This is reflected in the sustained high temperatures of equatorial regions. It is not perhaps surprising that the average daily range of temperature in the area bounded approximately by Cancer and Capricorn exceeds the annual range, whereas in the non-tropical regions the reverse is true (Petterssen, 1958). The tropics are regions not only of high but also of fairly constant temperatures. Particularly in moist lowland regions, plant metabolism follows a pattern of almost uninterrupted activity.

Long-term change of climate

Vegetation maps of the world have no permanent validity. Change in the surface temperatures of the earth, such as occurred during the Pleistocene glaciations beginning about 1 000 000 years ago and ending within the last 20 000 years, would contract the area of lowland tropical vegetation. The progeny of temperate and circumpolar plants would

Figure 2 The maximum and minimum angle of the sun's rays at latitude 52° N and at the equator. Note that at the higher latitude the sun's heat is dissipated over a wider area.

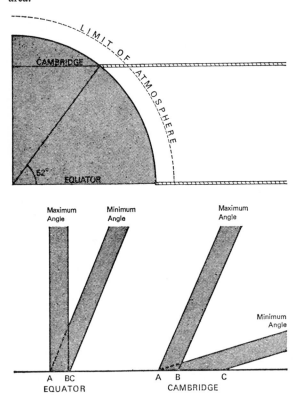

have to move towards the equator to survive under these conditions. While it is likely that glaciation exercised some effect in near-equatorial latitudes, obviously no exact data is available. According to Brooks (1949), equatorial rainfall would have been greater than at present and over the oceans trade winds would carry more moist air due to the greater contrast between polar and tropical temperatures. Interestingly, Boughey (1957) suggests that equatorial Africa was subjected to some drier periods than other areas during the Pleistocene period. Another consequence of glaciation was the lowering of montane floristic regions and the snowline. This latter appears in the tropics to have been depressed between approximately 400 and 600 metres below its present level (Brooks, ibid).

During the long existence of our planet there have been other periods of glaciation, but as they recede in time they become increasingly difficult to study. The Pleistocene glaciations are of particular interest to botanists because they occurred after the elaboration and spread of the angiosperms, which, as a group, still bear the marks of that tremendous climatic event.

Glaciation involves, in general terms, the extension towards the equator of the polar icecaps and, as would be expected, evidence of ice movement is found at comparable latitudes in both the northern and southern hemispheres.

Continental drift

Periodic cooling of the earth at remote intervals is not the only way in which a particular region may be exposed to differing climatic regimes. There is the possibility of a land mass drifting from one latitude to another—a pattern of events envisaged by the theory of continental drift which may be explained as follows. The approximate similarity in outline of the east coast of South America and the west coast of Africa has long been apparent and the notion that they were once joined together has a long history. In recent times Wegener advocated this and suggested that most of the world's land mass was originally one huge continent. This Wegener called Pangaea. Gradually this may have fragmented, as shown in figure 3. For many years argument has gone on about the validity of Wegener's views and a substantial body of opinion is coming to accept a modified theory of continental drift. While it is, for example, apparent that a very close structural correlation occurs between the supposedly once-joined coasts of America and Africa and for other similar regions too, the most recent and convincing evidence has come from a new field, 'palaeomagnetism'. Stated very simply, this refers to the property of molten ferruginous or iron-containing rock becoming magnetised as it cools relative to the earth's magnetic field and retaining through subsequent events its original magnetic axis. Thus if the position of such rock were to change, as it might if the land mass were drifting on a curving course, the axis of magnetisation would no longer be aligned to that of the earth's magnetic field. It forms what is very aptly described as a 'fossil compass'. By examining many such 'compasses' and making due correction for the date of magnetisation it is becoming increasingly evident that non-alignment to the earth's magnetic axis can be explained by assuming continental movement of a kind suggested by other evidence.

While even a modified theory of continental drift is not universally

Figure 3 Wegener's reconstruction of the continents during the periods indicated. Lined, ocean; dotted, shallow seas; latitude and longitude arbitrary.

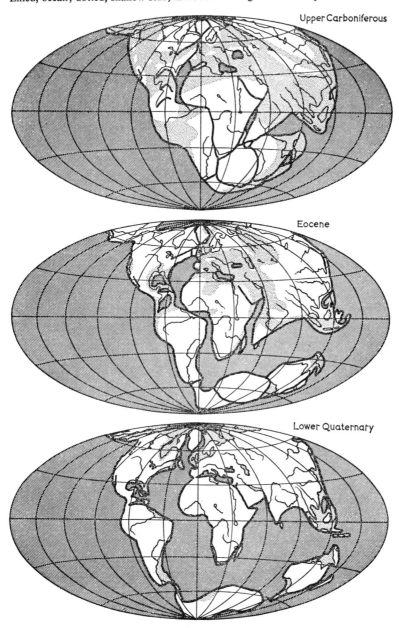

Upper Carboniferous

Eocene

Lower Quaternary

accepted, it is none the less an important concept for plant geography. If one assumes that the large land mass started to fragment at a time when the angiosperms began to diversify this would explain both the world-wide distribution of flowering plants and the development of characteristic regional floras. The time of angiosperm emergence has been variously estimated from the Permian (Eames, 1961) to the Jurassic (Simpson, 1953). Du Toit (1957), in a review of the evidence for continental drift, considers trends initiated in the Jurassic produced appreciable drifting during Cretaceous times. If, as the theory requires, for instance, southern South America, South Africa and Australia were once joined together this goes some way towards explaining in all three places the occurrence of the families Cunoniaceae and Restionaceae. It would provide too one explanation for the presence of Oxalidaceae in both South America and Africa.

Man and the tropics

The traditional shifting cultivation of the tropics normally exerts little lasting effect on the composition of the flora or the structure of the habitat. When the cultivator moves on, the vegetation assumes its previous condition. Because the world's population is now rapidly increasing, casual cultivation is being replaced by the more efficient settled farming. This involves large-scale clearing and drainage, often with the added activities of terracing, levelling and irrigation, all of which have pronounced ecological effects.

The insistent rise of human numbers means that populations of other species will be altered too. What Man regards as useless will be swept aside, while his crop plants must be evolved towards new levels of productivity. Because of the sustained activity of plants in the moist tropical regions of the world we must increasingly turn towards these areas as a source of urgently needed food.

Tropical diversity

It will be evident that within the tropics there exists immense ecological diversity. There are extremes of heat and cold, of dryness and humidity. Some areas present an appearance of incessant growth and decay, while others show only sudden and short-lived activity. From a study of fossils and other evidence can be inferred climatic changes in remote geological time. From our knowledge of the world today we recognise that the face of the tropics is at present being transformed more rapidly than ever before.

Chapter 2

Plant variation

Many of our ideas about the plant kingdom are based upon the 'type' concept. This means that an individual plant, newly recognised as distinct, is described formally and named as a new species (or subspecies or variety). Subsequently every plant which resembles it sufficiently closely is referred to by the same name. The type specimen which was once growing in some obscure habitat thus acquires for the botanist a special significance, since it now has a peculiar relevance to all other preserved, living and even unborn generations of the same kind. In practice, a taxonomist will seek to establish that type material is properly representative of the species with which he is concerned. His published description will take account of the range of variation observed.

There is much to be said for this approach. Preserved specimens can be retained almost indefinitely for comparison. Identification, although intricate, is a rational process, and the conventional way of classifying plants provides a valuable documentation of the enormous number of plant forms. The extent to which other botanists interested in physiology, genetics or ecology automatically assume that the plant they are studying 'must have a name' and so often does, is a significant comment on the utility of taxonomy.

A difficulty with the type concept is that, in practice, a plant collected in the field seldom exactly resembles the type specimen. This can mean that either it resembles no type closely or that it is similar to one type in most but not all details. To such situations there are several possible responses, of which one is to regard the second specimen as a new type.

It is to some extent a matter of personal judgment, but clearly a tax-onomist should not increase the number of types unnecessarily nor, more importantly, misrepresent the biological relationship between the second specimen and the type. Especially with regard to the latter, taxonomy has had to become experimental and in so doing has been linked with genetics and ecology in the study of plant variation.

The causes of variation

Several plants referred to one species may differ structurally or in other ways due to either of the following causes:

(i) differences in environment may have influenced their developmen-tal processes sufficiently to cause dissimilarities;

(ii) alternative or additional genetic characters may be operating.

Without carefully controlled experiments it is usually difficult to attribute differences among plants of the same species to just one of these causes. Environmental and genetic changes can occur together and phase changes, such as the transition from juvenile to adult forms, sometimes provide an added complication. Before outlining some environmental and genetic effects it is necessary to define the terms 'genotype' and 'phenotype' originally adopted by Johannsen. The geno-type refers to hereditary material handed on to the offspring by its parents, whereas the phenotype is the individual produced when a given genotype develops in a particular environment.

Environmental variation

It is readily apparent that the luxuriance and productivity of a plant depend to a large extent on whether or not its environment is congenial. In spite of beliefs, for example, that the coconut tree does not need fer-tiliser, its rate of growth and yield are demonstrably improved by the addition of mineral nutrients.

Where a plant reproduces entirely vegetatively it is possible to describe more precisely the effect of varying the environment. In the case of edible bananas, which reproduce by suckers, replicates of the same genotype can be exposed to a range of different conditions. It has been observed that the time from planting to fruiting varies with alti-tude above sea level in Jamaica. For the variety Lacatan those planted at or near sea level fruit in about twelve months, while at 1200 metres it is nearer fifteen months (Tai, 1958). It is important to realise that while the genotype remained the same (neglecting for the moment the infrequent occurrence of mutations), this does not mean, necessarily, that in each environment the various genes functioned with the same relative effectiveness. The situation described for the banana is rather unusual, and among seed plants the genotype is normally also variable.

Apart from the more obvious effects of temperature, humidity, illumination and nutrient supply, the phenotype of a plant can be modified by its pathogens. Rice is sometimes infected by *Gibberella fujikuroi*, the imperfect stage of *Fusarium moniliforme*. An early symptom of the disease is that infected plants become pale and their stalks and leaves elongate, making them noticeably taller than healthy plants of similar age (Brian, 1959). Another example is that of the normally semi-prostrate *Euphorbia hirta*. Diseased plants infected with *Uromyces euphorbiae* subsequently adopt an erect habit (see plate 1).

Genetical variation

Genetical variation is that which can be inherited, and a simple instance is provided by the inheritance of flower colour in the periwinkle, *Catharanthus roseus* (formerly *Vinca rosea*). Flory (1944) found that pink-flowered plants crossed with white gave all pink-flowered F_1 progenies. The F_1 when selfed gave in F_2 approximately 9 pink: 3 white petals with red eyes: 4 all white, the new class being the two-colour one (see plate 2). Flory also observed that the red pigment in the stem was associated with pink flowers. Not all genetical variation is, however, so clear-cut as in the foregoing example. If a character is under the control of many genes or is subject to strong cytoplasmic influence, the pattern of inheritance is more complex. For the present it is sufficient to note the following properties of genetically controlled variation:

(i) the particular character shows a relatively uniform expression in a range of environments;

(ii) where there is a segregation of different genotypes they will occur in predictable proportions in a range of environments provided the genetic composition and breeding system of the parents are known and the environment is equally favourable to all variants.

Phase changes

Many plants show a markedly different appearance at various times of their life apart from the usual effects of growth. The recurrent leaflessness of some monsoon forest species is presumably an adaptation to seasonal drought. The thorny juvenile phase of *Citrus* equally seems to have an obvious adaptive value. Less easy to account for is the phase change of, say, *Marcgravia brownei*, a climbing plant of montane forest. This species ascends vertically a supporting trunk by means of a closely appressed stem with small leaves. Eventually fertile branches are produced hanging out from the stem and bearing both flowers and larger leaves (see plate 3). The adaptive significance of such leaf differences (heterophylly) remains unexplained.

Another interesting case is that of *Dalbergia brownei*. This plant has three stages, which are the 'juvenile' stage, the 'common mature' and the 'post mature' stages. The first is prostrate with compound leaves, while the second is a shrub with lax well-developed, sometimes twining branches and unifoliate leaves. The third stage is a high climber with much shorter lateral branches and unifoliate leaves (Adams, personal communication). In the Guttiferae an instance occurs where polymorphism seems to have caused taxonomic complication. *Calophyllum jaquinii* is a species known in both juvenile and flowering forms. *C. longifolium* is known only from sterile material closely resembling the juvenile stage of *C. jaquinii* (see Fawcett and Rendle, 1926). Less striking is the progressive simplification in leaf outline towards flowering of several *Ipomoea* species.

Some consequencies of variation

From whatever cause, it is evident that plant species are capable of variation, and this is important to the taxonomist, the plant breeder and the student of evolution. The taxonomist recognises the existence of 'good' and 'bad' characters, the former having a narrow range of expression which can, as a result, be conveniently recognised. Good taxonomic characters are less subject to environmental influence.

Plant variation provides the breeder with choices. He can choose between more and less productive individuals or he can select those with superior disease resistance. It is necessary in making these selections to distinguish environmental from inherited variation, since only the latter can be genetically manipulated to give the desired progeny. This has led to the idea of 'heritability', which may be explained as follows. The total phenotype variance V_p is derived from the genetic variance V_g and the environmental variance V_e so that

$$V_p = V_g + V_e$$

Heritability (H) is an expression of the amount of phenotypic variance which can be inherited:

$$H = \frac{V_g}{V_p} \quad \text{or} \quad \frac{V_g}{V_g + V_e}$$

If the value of V_e is very low, then H approaches 100 per cent. A character with high heritability is one that can be inherited almost unchanged. Conversely, where the environmental contribution V_e is large, the heritability value is low and selection is relatively ineffective.[1] (In the expression $V_p = V_g + V_e$ the component V_g can be expanded to take account, mathematically, of various kinds of gene action such as additive effects, dominance and interaction but such is beyond the scope of this book.)

For the evolutionist, interest in all kinds of variation lies in the adaptive significance to the species possessing them and it is necessary to examine more closely the terms 'species' and 'adaptation'.

Species

No biological term will persist in wide use unless it has some intrinsic value, but it does not follow that there will always be exact agreement as to its meaning. The word 'species' has been retained because living things do not form a continuous spectrum of intergrading forms but fall into discrete groups and if the word 'species' were discarded there would be a pressing need for another to replace it. Whether one views these groups primarily as consisting of organisms that look alike or as members of an interbreeding community, the existence of *groups* is not in doubt. Moreover, when either criterion of grouping is used the same individuals will often tend to occur together. The biologist, it would seem, is wedded to the idea of 'species' and, although he may raise all manner of apparently serious objection, divorce remains a less satisfactory alternative.

The existence of variation accounts for one shortcoming of the type concept but since the various related individuals occurring together share a common, and to some extent predictable, inheritance they can be treated collectively as a population. Because all plants potentially able to interbreed may be prevented from doing so by spatial isolation or for some other reason, the 'population' and the 'species' will seldom

[1] For a discussion of this topic see, for example, D S Falconer, *An Introduction to Quantitative Inheritance*, 1960, Oliver and Boyd.

exactly coincide, except in the case of certain rare endemics. The normal pattern is for a species to be represented throughout its geographic range in a series of partially or completely isolated populations. Heslop Harrison (1953) says:

'It is this concept of the statistically predictable population composed of evanescent individuals all linked with each other genetically through ancestry and all potentially capable of becoming linked again through their descendants which has come to occupy a central position in modern evolutionary thought.'

Behind the common features of plants from a given species there are processes of recombination and reassortment resulting in a segregation of variant individuals, whose collective identity is maintained by the elimination, under natural selection, of the extremes. In no two species (the word is chosen deliberately) are the genetic processes quite the same. Some species vary widely, others, hardly at all, and such variation as exists may be distributed slowly or rapidly among the members and it is important to realise that not all the variants of a species contribute equally to its continued survival.

Species are often divided into smaller parts which may reflect distinctive morphological or ecological units. Species, too, are grouped in progressively larger associations of genera and families and so on. None the less, for the reasons mentioned and for those in chapter 6, the species level is generally of most evolutionary interest.

Adaptation

Because the words 'adaptation' and 'adaptability' when applied to plants can have various meanings, it is essential to clarify certain ideas.

Perhaps the most instructive way to explain adaptation is to begin with a plant population. Normally this consists of a group of individuals that are related but not identical in either heredity or appearance. In such cases one can demonstrate that there are many approximately similar plants and rather fewer exceptional ones. This kind of population, if it continues to hold its own ecologically, is said to show adaptation to its environment. The overall adaptation of the population is the sum of several other kinds of adaptation, each of which can be identified.

No plant environment remains exactly constant and a useful distinction can be made between those changes that are either longer or shorter than the generation time of the individual plants being considered. If in the face of long-term change the population is to maintain its occupancy of the habitat continuing adaptation must occur. Where this happens the population demonstrates adaptability, and the proportions of various individual genotypes comprising it may change appreciably.

Turning now to the individual plant, it may, provided it lives long enough, persist through quite wide changes in its surroundings to complete the processes of growth and reproduction in a way that enables it to contribute either a larger or smaller number of surviving progeny to the population. If the former is the case, then this plant shows a high degree of individual *genotypic* adaptation. If the progeny are relatively uniform they represent a contribution to the 'fitness' of the population, whereas if they are genetically dissimilar they contribute to its 'flexibility'.

A particular genotype may be multiplied unchanged because it is clonally produced, or be almost unchanged if it is an inbred line that has become virtually homozygous. It thus becomes possible to put a particular genotype in several environments, and in each a corresponding phenotype will develop. If the environments are contrasting, then the successful genotype demonstrates 'phenotypic flexibility', which may express itself in two ways. Either in each environment the plant will show a similar morphology or in each case distinctive phenotypes may develop. The former response has been called 'developmental homeostasis' and the latter 'plasticity'. (The terms 'homeostasis' and 'plasticity' have been used in various ways by different authors and in different contexts would have other meanings.) 'Adaptability' is sometimes applied to individual genotypes, but it must be remembered that the genotype of an individual is decided at syngamy and, apart from mutation, cannot change in higher plants. An individual genotype is not adaptable in the same sense as a population (or at least a line of descent) which in response to selection may undergo sustained genetic change.

Even in the environment most conducive to a plant's continued survival, it is unlikely that all its metabolic processes will run continuously at optimum. Where a species' range becomes consistently extended away from its most congenial environment, the need for adjustment to or toleration of a sub-optimal metabolism is presented with increased insistence. Whether the individual response is plastic or homeostatic, this provides a 'toehold' mechanism until such times as the resulting progeny segregate a less precariously adapted genotype—a situation described as the Baldwin effect after its first proponent.

Variation and survival

From what has been said it will be seen that while, traditionally, an individual plant has been regarded as a member of a species, it is more instructive to regard it as part of a population. Not only is it a sample of the genetic material available among a group of related individuals but also it is a potential contributor to the make-up of succeeding generations. A partial analogy would be that of regarding individuals as figures in a kaleidoscope. The genetic material is repeatedly rearranged to create continuing variation. The genetic kaleidoscope, because of linkage, turns gently at each generation and figures tend to persist with only small changes. This analogy breaks down because natural selection and mutation through time change the pieces from which the figures are made.

The continued survival of a population depends on its adaptability and competitiveness through successive generations. This in turn is an expression of the phenotypic flexibility of individuals and the genetic properties segregated to their progeny.

Chapter 3

Modification of the flower

A functioning flower ensures production of fruit and seed, and usually both male and female organs (stamens and carpels) are present. Around them are other structures in some degree conspicuous or protective. Such a flower represents both an important product of evolutionary activity and the starting-point for several more. It has been suggested that the primitive angiosperm flower was derived from an hermaphrodite strobilus consisting of a spiral of megasporophylls above a spiral of microsporophylls. The latter are considered to have become progressively sterilised to form petals or retained their fertility to become stamens. The megasporophylls eventually evolved into carpels. There is now considerable support for this view, and among present-day plants *Magnolia* and some of its woody relatives are regarded as relatively primitive in view of their floral structures.

Among the angiosperms the dependence of both male and female gametophytes upon the sporophyte is extreme. Frequently they are confined to a common strobilus. Such an arrangement has two important effects, one seemingly convenient and the other less so. The production of gametes creates a break in the life cycle which imposes a risk that is only overcome at gametic fusion. The close proximity of the stamens and carpels certainly increases the likelihood of gametes meeting but it severely limits the prospects for crossing. Chance variations of whatever kind which increased the possibility of crossing without appreciably diminishing the overall fertility would have both conferred advantages on their possessors and also initiated new floral trends.

Figure 4 Floral changes which promote emphasis towards inbreeding or out-breeding. The suppression of one or other sex organ, a departure from simultaneous functioning of male and female parts or the interposing of an incompatibility mechanism renders self-pollination less likely and favours crossing. The converse, where to simultaneous functioning is added substantial exclusion of other pollen, promotes selfing and consequently inbreeding.

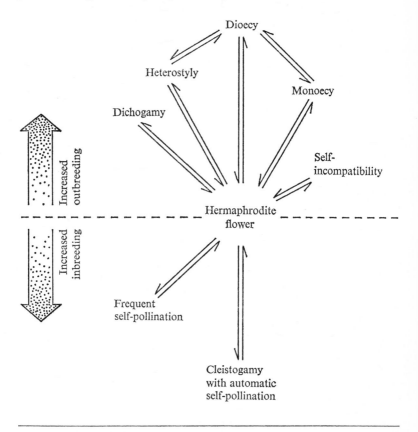

Plate 1 *Euphorbia hirta* showing normal semi-prostrate stems and erect ones which are infected with *Uromyces euphorbiae*. Pustules of the rust are evident on the abaxial leaf surfaces of erect stems.

Plate 2 Segregation in F_2 of *Catharanthus roseus*. W (pink petal), w (white petal), R (red eye), r (white eye). Crossing pink petal, red eye with white petal, white eye gives F_1 $WwRr$ (pink petal, red eye). The F_2 generation yields among the various recombinations those which show homozygous recessive petal colour combined with one or more dominants for eye colour ($1wwRR, 2wwRr$). This is the new class appearing at F_2.

Plate 1

Plate 2

Plate 3

Plate 4

To argue in this way about angiosperm evolution assumes that cross-fertilisation is beneficial. As early as 1876 Darwin had shown that prolonged self-fertilisation depressed the vigour of offspring while crossing could restore and maintain it. Fisher (1958) pointed out that heterozygotes (which tend to be produced by crossing) can segregate more than one kind of genotype and consequently have more adaptive possibilities. Cross-fertilisation, therefore, has two important characteristics. It is associated with greater vigour and the progeny of resulting crosses tends to be genetically diverse.

A higher frequency of heterozygotes can be achieved by outbreeding. This is made possible by hindering selfing and encouraging crossing, although the first of these has no survival value unless it is accompanied by the latter. Figure 4 shows various ways in which self-pollination is hindered or prevented and how it can be accentuated.

In figure 4 various means of affecting breeding behaviour are shown and it is now appropriate to make the following points, namely:

(i) most of the restraints upon selfing are neither irreversible nor absolute;

(ii) the possibility exists of one anti-selfing mechanism being transformed to another;

(iii) it is possible for one individual to carry more than one anti-selfing mechanism simultaneously;

(iv) where a species has appreciable geographical distribution, different systems to prevent selfing may operate in various parts of the range;

(v) restraints upon selfing need not affect all the flowers of one individual or all the members of a population.

Since each of these is important, they will be discussed in turn.

Self-incompatibility

Nicotiana sanderae is self-incompatible. Several alleles occur at one

---

Plate 3 Heterophylly in *Marcgravia brownei*. The smaller appressed leaves are on an ascending stem on the tree, while the larger leaves are on the lateral branch to the left.

Plate 4 Automatic interspecific hybridisation in *Croton*. To the left is *Croton linearis* and to the right *C. flavens*, while lower centre is *C. linearis* × *flavens*. Note that the adaxial leaf surface of the hybrid is intermediate in texture between the glossy female and glandular hairy male parent.

B

locus and pollen carrying a particular allele cannot send its tube through style tissue carrying the same allele. Figure 5 shows the degree of success with which various combinations of alleles can be crossed. Reciprocal crosses behave similarly.

Figure 5 Incompatibility in *Nicotiana*.

Female diploid tissue	Female haploid tissue	Pollen S_1 or S_2	
S_1S_2	S_1 or S_2	——	——
		——	——
S_1S_3	S_1 or S_3	——	$\boxed{S_1S_2}$
		——	$\boxed{S_2S_3}$
S_3S_4	S_3 or S_4	$\boxed{S_1S_3}$	$\boxed{S_2S_3}$
		$\boxed{S_1S_4}$	$\boxed{S_2S_4}$

(Meiosis)

—— = no pollen tube growth

▢ = diploid genotype of new zygote

Note: A pollen tube will germinate where its allele is not common to either one in the stylar tissue.

In the first case neither pollen grain can function since each has an allele in common with the stylar tissue, while in the second case half the pollen grains lack an allele in common with the style and can function. The third case shows a situation where pollen and style tissue share no alleles and the pollen grains are all, theoretically, functional.

In addition to those mentioned a self-fertility allele S_f also occurs. Where it does, individual plants can be successfully self-pollinated. S_x is any self-incompatibility allele. The combination S_fS_x can be selfed to give S_fS_f and S_fS_x individuals.

A considerable amount of literature now exists about incompatibility. What appear to be quite similar situations show on closer analysis all manner of subtle difference. A novel 'sporophytic' incompatibility occurs in cocoa (*Theobroma cacao*), where the effectiveness of an incompatibility allele depends in some degree on the sporophyte from which it arises (Cope, 1962).

Transformation

Jones (1934), using carefully selected recessive genes, converted mono-ecious maize into one reproducing dioeciously. He used the gene *sk*, which suppressed growth of silks (styles), and the gene *ts*, which gives rise to seeds in the tassels (male flowers). With these the following crosses were made:

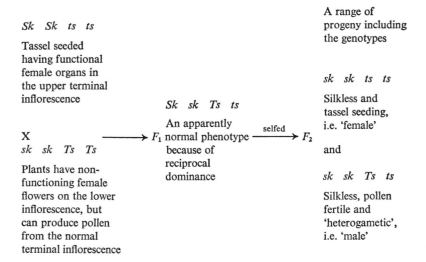

Sk   Sk   ts   ts

Tassel seeded
having functional
female organs in
the upper terminal
inflorescence

X

sk   sk   Ts   Ts

Plants have non-
functioning female
flowers on the lower
inflorescence, but
can produce pollen
from the normal
terminal inflorescence

Sk   sk   Ts   ts

An apparently
F_1 normal phenotype $\xrightarrow{\text{selfed}}$ F_2
because of
reciprocal
dominance

A range of
progeny including
the genotypes

sk   sk   ts   ts

Silkless and
tassel seeding,
i.e. 'female'

and

sk   sk   Ts   ts

Silkless, pollen
fertile and
'heterogametic',
i.e. 'male'

If the two F_2 genotypes shown are crossed they will continue to reproduce themselves in equal proportions in the progeny.

Other examples of reversibility can occur.

The tomato (*Lycopersicon esculentum*) is normally cross-pollinated by small halictine bees. Under glasshouse conditions the pollen vectors are absent and automatically self-pollinated variants where the stigma does not protrude through the anther cone have survived preferentially. In the genus *Solanum* a self-incompatibility mechanism operates in the diploid species. Conversion to tetraploids such as occurred in nature caused the resulting plants to become self-compatible and the chance occurrence of diploid reversions restores the effectiveness of the self-compatibility system (see chapter 7).

Dual anti-selfing mechanisms

Red clover (*Trifolium pratense*) is both protandrous—an example of dichogamy—and self-incompatible. Perhaps the most curious case is that of *Rhamnus cathartica*, where dioecy seems to have been super-imposed upon heterostyly, without erasing its expression (Darlington, 1958: see figure 6).

Figure 6 Dioecy and heterostyly in *Rhamnus carthartica*. Left, long-styled; right, short-styled.

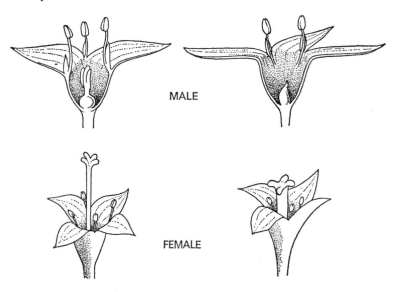

MALE

FEMALE

Geographical variation

Any character possessing an appreciable degree of heritability is subject to selection, especially if it affects the adaptability of its possessor. What may be an advantage in one location need not be so in every other that the species has colonised and this will be so for floral structure as for any other feature. Taylor (1954) claims that *Coprosma pumila* is dioecious in New Zealand and monoecious on Macquarie Island. Stebbins (1957) provided other examples such as occur within the genus *Secale*. *S. montanum* occurs in the mountains from Morocco to Iran while *S. africanum* is an isolated species (or perhaps ecotype of *S. montanum*). *S. montanum* is self-incompatible while *S. africanum* is quite self-compatible. In 1955 Baker pointed out that a plant derived from a single propagule accidentally dispersed over a long distance cannot establish a colony unless it is self-fertilising even though it may not have been in its previous or original habitat.

Selfing restraints and individual plants

The production of non-selfing flowers need not involve the entire inflorescence of one plant nor the whole population within which they occur. The mango, *Mangifera indica*, produces inflorescences with both male and hermaphrodite flowers and many other examples can be found. Apart from strictly dioecious plants such as the nutmeg, *Myristica fragrans*, there exist some that combine dioecy with other floral systems. The paw paw, *Carica papaya*, for example, has populations that include males, females and hermaphrodites and the breeding system that ensures this is well worked out (see figure 7).

Figure 7 Inheritance of sex in paw paw. There is one gene with three alleles, M_1, M_2 and m. All combinations of dominants M_1M_1, M_1M_2 and M_2M_2 are lethal (underlined here). M_1m is male, M_2m is hermaphrodite and mm female. (Storey, 1957.)

(1) $mm \times M_1m \longrightarrow 1\ mm: 1\ M_1m$

(2) $mm \times M_2m \longrightarrow 1\ mm: 1\ M_2m$

(3) M_1m selfed and $M_1m \times M_1m \longrightarrow 1\ mm: 2\ M_1m: 1\ \underline{M_1M_1}$

(4) M_2m selfed, and $M_2m \times M_2m \longrightarrow 1\ mm: 2\ M_2m: 1\ \underline{M_2M_2}$

(5) $M_2m \times M_1m$, and $M_1m \times M_2m \longrightarrow 1\ mm: 1\ M_2m: 1\ M_1m: 1\ \underline{M_1M_2}$

Another tree having an unusual floral development is the aceituno *Simarouba glauca* (Armour, 1959). Individual trees have one of three kinds of flower development, a pistillate form, a staminate form and an andromonoecious form where trees have both hermaphrodite and male flowers.

Although the literature contains such words as gynomonoecious (having hermaphrodite and female flowers) and gynodioecious (having some plants hermaphrodite and others female), the verbal inventiveness of botanists should not obscure the principal fact that different arrangements of perfect and imperfect flowers reduce selfing, especially if reinforced by a measure of dichogamy.

It remains to draw attention to an interesting exception in the foregoing work. Apart from the paw paw, all the various modifications seem to have developed along so-called 'classical' lines, which is to say that where an hermaphrodite flower becomes unisexual it does so because of the atrophy of either the gynoecium or the androecium. Some work of Storey (personal communication) shows that the paw paw is exceptional in this respect. While the male flower was derived classically by the phylogenetic loss of carpels, transmutation of stamens into carpels (carpellody) in the hermaphrodite tree can take place. This is a seasonal occurrence more common in certain strains and leading to undesirable 'cat-faced' fruits.

Angiosperm diversity

Evolutionary development from the original angiosperm stock has clearly been on an impressive scale when one considers the range of vegetative forms, habitats colonised and array of floral structures.

The exact routes by which advanced species evolved from more primitive ones are still subject to speculation, although there is no reason to believe that anything other than normal genetic processes has provided the mechanisms of change. In some ways our experimental techniques are limited and it is, for example, not always possible to cross distantly related plants and collect data from the ensuing hybrids and segregants. A great deal, however, is possible experimentally and modification of the flower along 'taxonomic' lines can be readily demonstrated. In the Scrophulariaceae, for example, genes occur capable of converting zygomorphic corollas to actinomorphic ones, as in *Calceolaria*. The alteration by Jones of the maize breeding system has already been mentioned and, since this topic has important implications, it is treated in greater detail in chapter 9.

The diversity shown by angiosperms has its greatest profusion in the number of floral forms which occur. The flower, it may be noted, often shows co-adaptation with more than one kind of animal. To take a common case, the petal colour and placement of pollen and stigma can be well adapted to bee visits, while the ensuing fruit or seed dispersal depends upon the food preferences of quite different animals. Not all animals are so benign. Flowers and fruits are subject to numerous specifically adapted predators. These, too, make demands upon the adaptability of plant species.

Floral structures became established, in the case of entomophilous plants, because they interacted successfully with insect pollinators. Since the means of pollen transfer are so varied in the tropics, the next chapter is concerned with this subject.

Chapter 4

Pollination

Since plants that result from cross-pollination tend to have both greater vigour and more genetic diversity than those arising from self-pollination, it is not surprising that the former are more common. This is true whether one considers numbers of species or numbers of individuals. Separation of the sexes (either spatially or by some other means) hinders self-pollination but this would only be a disadvantage if there were no means of pollen transfer between individuals. Reliable pollen transfer could scarcely arise at once. We may conceive, therefore, that the mutual dependence[1] evident today between some entomophilous flowers and their insect visitors was preceded by much more casual associations of unspecialised flowers and relatively undiscriminating visitors. From such beginnings one can envisage a series of events involving specialisation in floral development and the increasing exclusion of those insects less well adapted to the flower as a result. Such a sequence would be unlikely to persist indefinitely without interruption, deviation or retrogression because different pollinators might have competed for the same floral products or deserted them almost entirely from time to time. The available evidence suggests that the earliest animal pollinators were possibly beetles and that only later did bees, moths and butterflies begin to achieve their present prominence.

---

[1] Since this book was written, a comprehensive review of this subject has been published by H G Baker and P D Hurd: 'Intra-floral ecology', *Annual Review of Entomology* 1968 (see bibliography).

Because the populations of plant species depend so largely on their pollinators, it is necessary to examine aspects of pollen transfer.

Beetles as primitive pollinators

If beetles were among the first pollen vectors we might hope to find them active today as pollinators of what are regarded as primitive angiosperms. We might also expect fossil relatives of these beetles to antedate the earliest fossils of, say, bees and bee flowers. Although the biology of temperate flowers is better known, as early as 1883 Muller recognised that in the tropics and subtropics beetles were more strikingly adapted to a floral diet. The pollination of the spice bush *Calycanthus occidentalis* (Calycanthaceae), described by Grant (1950), shows that *Colopteris* beetles pollinate the flowers if they are available in the area, but if not they seek other sources of forage.

Moths and butterflies

Both moths and butterflies are important as pollinators since they extract nectar from many species of plants. Their host plants, typically, have either long tubular flowers or flowers with tubular nectaries.

Probably the most famous instance of moth pollination is that of the Madagascar orchid *Angraecum sesquipedale*. Wallace (1891), having observed the long tubular nectary of this plant, wrote of its then unknown pollinator:

'That such a moth exists in Madagascar may be safely predicted; and naturalists who visit that island should search for it with as much confidence as astronomers searched for the planet Neptune—and I venture to predict they will be equally successful!'

Darwin in 1862, equally confident though less flamboyant, wrote:[2]

'In Madagascar *Angraecum sesquipedale* must depend on some gigantic moth.'

Subsequently a moth, *Xanthopan morgani* f. *praedicta*, with the necessary attributes was described.

Bees

Bees collect both nectar and pollen assiduously and in fact base almost their entire economy on the substances produced by flowers. The bee that is most familiar, the honeybee, *Apis mellifera*, is in some ways a

[2] See bibliography.

very unusual species. Its almost world-wide range, elaborate social structure, sophisticated navigation, ease of domestication and varied choice of forage are, taken together, equalled by no other bee. Conclusions drawn, therefore, from honeybee studies do not necessarily apply to other bee species. To take but one example, a social stingless bee species, although having a somewhat advanced colony structure, navigates partly by means of scent marks and is thus prevented from guiding other bees across quite small bodies of water (Lindauer and Kerr, 1960).

Among the forage habits of bees one can distinguish 'polytropic' and 'oligotropic' types, meaning that many or few species of plants respectively are visited. The terms 'polylectic' and 'oligolectic' refer to the relative numbers of species, many or few, visited for pollen. The worker honeybee, although polylectic, will more often than not confine herself mostly to one species during a particular flight from and to the hive. At different times of the same day or on subsequent days the same bee will transfer her attention to other pollen sources if the original supply begins to fail. Polylecty does not imply non-discrimination.

A contrast to the honeybee in foraging behaviour is provided by the oligolectic squash bees belonging to the genera *Peponapis* and *Xenoglossa*. These bees collect pollen exclusively from pumpkins, squashes and gourds (*Cucurbitaceae*), (Michelbacher, Smith and Hurd, 1964).

Other pollinators

In addition to those already mentioned, other animals such as bats, birds and ants are involved regularly in pollination and the first will be discussed subsequently. Although these animals participate in unusual pollination devices it is generally recognised that such mechanisms are derived, secondarily, from more conventional ones and involve only a minority of flowering plants.

Flowers and their pollinators

In general, neither the pollinator nor the plant has a monospecific allegiance to one partner. There are well-authenticated cases of extreme fidelity by one partner, but this seldom seems entirely reciprocated. Although *Peponapis* confines itself to certain plants of the Cucurbitaceae, these are frequently visited by the honeybee, for example. Conversely, monkshood, *Aconitum*, is pollinated only by humblebees although such bees have a varied diet and visit many genera.

The most extreme instance of mutual dependence between a flower and a pollinator involves not a bee but a wasp, *Blastophaga psenes*, and the fig, *Ficus carica*.[3]

The fig–wasp relationship is exceptional and the most common condition is that of a plant species visited by several types of pollinator at any one place and where one or two types will be relatively common. For the same plant species in another area a similar pattern will occur but representation among the vectors will be altered. It is quite possible that in any one place the range of pollinators will be wide. *Pentas lanceolata*, for example, in Jamaica is worked regularly by many butterflies and a humming bird for nectar and for both nectar and pollen by several species of bee.

The evolution of insect pollinator to flower relationship must obviously have involved modifications of host and visitor. If this process were to continue on a large scale a situation could arise where there are so many 'locks' (floral modifications) and a correspondingly large number of 'keys' (pollen vectors) that a particular key might open more than one lock even though it may work in the 'wrong' way. Something like this seems to have happened with two tropical orchids, *Gongora maculata* and *Coryanthes speciosa*. The pollination mechanism of each is complex and quite different in the two cases.

'This orchid [*Coryanthes*] has part of its labellum or lower lip hollowed out into a great bucket, into which drops of almost pure water continually fall from two secreting horns which stand above it; and when the bucket is half full, the water overflows by a spout on one side. The basal part of the labellum stands over the bucket, and is itself hollowed out into a sort of chamber with two lateral entrances; within this chamber there are curious fleshy ridges. The most ingenious man, if he had not witnessed what takes place, could never have imagined what purpose all these parts serve. But Dr Cruger saw crowds of large humble-bees visiting the gigantic flowers of this orchid, not in order to suck nectar, but to gnaw off the ridges within the chamber above the bucket; in doing this they frequently pushed each other into the bucket, and their wings being thus wetted they could not fly away, but were compelled to crawl out through the passage formed by the spout or overflow. Dr Cruger saw a "continual procession" of bees thus crawling out of their involuntary bath. The passage is narrow, and is roofed over by the column, so that a bee, in forcing its way out, first rubs its back against the viscid stigma and then against the viscid glands of the pollen-masses. The pollen-masses are thus glued to the back of the bee which first happens to crawl out through the passage of a lately expanded flower, and are thus carried away.'

Darwin, 1859

[3] The reproductive biology of the fig is not simple. Parthenocarpy occurs and has been exploited by Man in the cultivated fig. Other fig species have mutually dependent relationships with particular wasp species.

'The male bee is attracted to a flower (of *Gongora*) by its fragrance, the source of which is a deep cleft at the base of the lip. If in attempting to reach this the bee clambers onto the keel-like, slippery plates of the lip it may fall, slide on its back over the smooth concave surface of the column, strike the tip of the anther and remove the pair of pollinia. Insertion of these pollinia from the back of the insect into the very narrow stigmatic cleft is possible only after a rather protracted period and in this way cross-pollination is more likely to occur than selfing.'

Baker, 1963

Interestingly, the same bee species, *Euglossa cordata*, operates both mechanisms. From this Baker (1963) concludes that one or both of these systems is of recent sudden evolution. In these pollination mechanisms it is the male bees that are attracted by scent to the flowers. Baker considers that adaptive change in the two orchids without change in the bee is possible, since the male insect visits involve the bee species in no disadvantage, there being, as he says, almost no other calls on their time.

Discrimination

In 1876 Darwin wrote:

'Humble and hive bees are good botanists for they know that varieties may differ widely in the colour of their flowers and yet belong to the same species.'

Since Darwin's time the taxonomic discernment of bees has been examined particularly by Mather (1947) using *Antirrhinum*, Grant (1949) using *Gilia* and Bateman (1951) working with *Brassica*. While all three of these investigations are most interesting, specific reference will be made here only to that of Bateman. This worker used *B. oleracea* (cabbage) and *B. napus* (swede). The latter is an allotetraploid hybrid of *B. oleracea* and *B. campestris* (turnip). Two varieties of cabbage and two of swede were used. Hive bees visited both species but appeared to adopt one or other on a particular flight. They also showed a lower, but none the less significant, constancy to a particular variety. Bumblebees visited only cabbage, while solitary bees visited only swede.

Strictly tropical experimental data on insect pollinator discrimination is almost non-existent in spite of the wealth of opportunities that exist for gathering such information. Some work, however, of Dronamraju and Spurway (1960) may be quoted, involving several species of butterfly on pink and orange horticultural varieties of *Lantana camara* (Verbenaceae). They showed that there were colour preferences towards one or other variety (see table 3) and in the case of *Papilio demoleus* gave evidence that the preference was innate.

TABLE 3. FLORAL PREFERENCES OF SIX SPECIES OF BUTTERFLY
TOWARDS FLOWERS OF *LANTANA CAMARA*

Species	Family	Date of 1st obsv.	Date of last obsv.	No. of days obsvd	No. of feeds on orange	No. of feeds on pink
Precis almana	Nymphalidae	29.5.58	21.4.59	16	218	13
Danais chrysippus	Danaidae	19.3.59	22.5.59	18	142	152
Papilio polytes	Papilionidae	19.3.59	4.4.59	4	15	31
Papilio demoleus	Papilionidae	19.3.59	19.5.59	13	42	98
Catopsilia pyranthe	Pieridae	29.5.58	22.5.59	27	40	603
Baoris mathias	Hesperiidae	19.3.59	19.5.59	12	1	108

Source: Dronamraju and Spurway, 1960

The present author working with *Pentas lanceolata* (Rubiaceae), a heterostyled ornamental, showed that individual honeybees had preferences for clones of similar colours and style length. The assiduity with which a particular clone was worked was correlated with the fullness and quantity of pollen grains. It is probably significant that throughout the wide range of colour variants in this species the stigma in the long-styled form is usually coloured similarly to the anthers in the colour-corresponding long-anthered form. The effect must be presumably to disguise the differences and minimise the risk of vector discrimination between style lengths—an essential if the heterostyled breeding mechanism is to succeed.

Since there is a range of floral differences from small to large to be found within a species, it is conceivable that some would be of just sufficient magnitude to disconcert a pollen vector. If this were so a pollinator would discriminate against the mutant, with the result that one continuous population could be subdivided into two discrete subpopulations. This would be 'isolation without distance'. The converse possibility is that representatives of two distinctive and separate populations of the same species might be brought together and share a pollen vector, with subsequent genetic merging if their differences were sufficiently minor. The first of these cases, 'isolation without distance' as a speciation mechanism, while theoretically possible, does not seem to have been unequivocally demonstrated (see, however, page 55). What is much more frequent is the situation where already existing differences between two races or subspecies arising from other causes are maintained or reinforced by the discrimination of their vectors.

It is however sadly true that the tropical environment with its myriads of pollinators, its dawn-to-dusk activity throughout virtually the entire year and its profusion of plant species remains short only of investigators prepared to tolerate the small discomforts that accompany waiting, watching and recording.

Bat pollination; a curious problem

The genus *Parkia* is adapted to pollination by bats. A West African species, *P. clappertoniana*, although visited by honeybees for nectar and pollen, seems more effectively pollinated by fruit-eating bats (Baker and Harris, 1957). The spherical inflorescence (see figure 8) has a rim which collects nectar and from which bats can lap. In so doing, pollen is placed on and rubbed off the fur of the bats' face and breast. Stomach contents of such bats included nectar but no remains of stamens and flowers to suggest chewing. Two species of bats, *Epomophorus gambianus* and *Nanonycteris veldkampii*, were recorded pollinating *Parkia* in West Africa by Baker and Harris. Both bat species belong to the section Megachiroptera, an exclusively Old World fruit-eating group.

The genus *Parkia* is of pan-tropical distribution, occurring in West Africa, the Amazon region of South America and the West Indies and in continental South-East Asia and the East Indies. Whether because of climatic shifts or by continental drift, the principal distribution of *Parkia* must have occurred before the end of the Eocene. In Africa and Asia *Parkia* is pollinated by the Megachiroptera, while in South

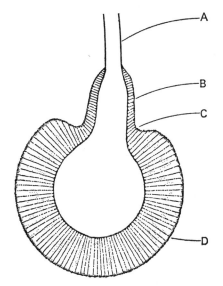

Figure 8 Diagrammatic section through the inflorescence of *Parkia clappertoniana*.

Based on Baker and Harris, 1957.

(A) stout peduncle;
(B) sterile nectar-producing flowers;
(C) ring in which nectar collects;
(D) potentially fertile flowers.

America it is visited by species of bats belonging to the *Microchiroptera*, *Anoura geoffroyi* and *Glossophaga soricina*.

The Microchiroptera are a group of largely insectivorous bats in which fruit feeding developed secondarily and which are confined to the New World. Megachiroptera are fruit feeding and confined to the Old World. It seems, therefore, that the distribution and survival of *Parkia* (already bat-adapted) preceded considerably the evolution of New World fruit-feeding bats. Bat fossils are unknown before the Eocene, so presumably either *Parkia* in the New World provides a curious case of pre-adaptation to bats which subsequently evolved there or the bat fossil record will eventually be shown to extend to earlier times.

Pollination in the tropics

This account, although necessarily brief, perhaps goes some way towards setting out a key area of evolutionary enquiry. As will be evident, the ground was broken here during the nineteenth century as in so many other fields. Most of the area of tropical pollination awaits a systematic and experimental study, especially one that gives prominence to bees.

Chapter 5

Gene expression and diploidy

Throughout the plant kingdom there is an obvious trend towards complexity coincident with the gradual rise to predominance of the sporophyte. Where the greater part of the life cycle is spent in the haploid condition, plants are generally fairly simple in structure. Among predominantly diploid organisms the morphology is more complex and more varied. Presumably, therefore, the diploid condition is more susceptible to this sort of elaboration. Why should this be?

Svedelius (1929) suggested that if the fusion of haploid gametes is not immediately followed by meiosis but by a series of mitoses a multi-cellular structure results, assuming the cells maintain contiguity. This could ultimately accommodate a larger number of meiotic divisions with the greater likelihood of better adapted genotypes. While this may partly explain the relatively larger sizes of diploid plants, one must not automatically equate size with success. Of the cryptogams, micro-organisms in particular appear to be simple, short-lived and opportunist when compared with higher plants. While all plants have the problem of exploiting such variation as occurs to best advantage, it does not follow that one method of contriving genetic change will be invariably superior. Among micro-organisms there are a variety of what Ponte-corvo (1954) has called 'parasexual' mechanisms. Angiosperms are elaborate and long-lived and their development is complex and closely integrated, one result of which has been that only a small proportion of the plant body is adapted to a recombining function.

Svedelius' suggestion provides one reason for the increased size of chiefly diploid plants, but implicit in it is the idea that segregation is a valuable or worthwhile process. Because segregation is (in the conventional sense) impossible in haploids but customary, if not universal, among diploids, it is instructive at this juncture to examine some of the consequences of segregation that diploidy permits.

Recombination

One result of sporophytic predominance is the altered role of meiosis. For mosses and liverworts the sporophyte is normally so transient as to seem at first to have the reinauguration of the gametophyte as its sole function. By contrast, among seed plants it is the gametophyte which has a brief existence, with its significance apparently restricted to that of ensuring a link between meiosis and syngamy. Among the bryophytes, therefore, the products of meiotic recombination are tried out in the haploid gametophyte stage immediately following, but in the angiosperms the products of meiosis are not appreciably exposed until the succeeding diplophase. This means that each chromosome (incorporating a set of genes) must function effectively not only with different chromosomes but also with its own homologue.

There are several factors that promote recombination—notably short generation times, high chromosome numbers, high chiasma frequency, outbreeding and few incompatibility barriers between related species (Grant, 1958). Linkage and the contrary situations to those just mentioned retard recombination. It follows, of course, that if all the factors making for high recombination values are operating and variability is limited, their effectiveness is diminished.

Dominance

Associated often but not invariably with diploidy is 'dominance'. This may be defined as the property shown by an allele at one locus to obscure the effects of other alleles at the same locus. Pink flowers in the periwinkle (see chapter 2) are dominant to white. The pink allele obscures the effect of the white one at the same locus. Any population in order to survive has to preserve a group of individuals and these individuals have to contain among themselves the variation that will be needed in succeeding generations. Three properties of dominance, therefore, are especially useful here. These are:

(i) since there is a gene-by-gene control of plant activity, where a recessive combination occurs for a particular organ or feature, other well-adapted organs continue to function normally or nearly normally;

(ii) a dominant allele, well adapted to a particular situation, can obscure the effect of various recessives, each somewhat different, that can be substituted at the same locus;

(iii) dominance itself is not immutable and under pressure of selection can be modified so that one allele hitherto recessive can supplant another as dominant for that locus (see below).

Heterozygosity

Dominance and recombination in a diploid permit the cross between genotypes $AA \times aa$ to yield the hybrid Aa. Selfing this or intercrossing Aa types gives three genotypes, AA, Aa and aa. They occur in predictable frequencies and the heterozygote Aa is the most common. Where the diplophase is the predominant part of the life cycle the dominant and recessive alleles coexist during the processes of growth, elaboration and maturity. (Where dominance does not occur the coexistence of alleles a_1 and a_2 in the hybrid a_1a_2 is implied.)

Homo- and heterozygotes thus differ in two respects. While the former can only reproduce true to type, the latter can segregate three kinds of progeny. This is the genetic aspect. The second difference is physiological. Since the differing alleles to some extent complement each other biochemically, the heterozygote will 'buffer' environmental harassments more effectively than homozygotes. The development of heterozygotes is less likely to be put out of order than is that of a homozygote.

A spectacular and extreme form of the physiological effect of interacting alleles occurs in hybrid corn. Dissimilar pure lines when hybridised can sometimes result in progenies that greatly outyield either parent. Whether the effect is promoted by 'complementary dominance' or by gene interaction of some other kind is uncertain but this has not prevented its exploitation in the development of commercial corn (see chapter 9).

Response to selection

The progeny from a selfed heterozygote will be AA, Aa and aa. If the recessive is then removed or prevented from reproducing, repeated selfing of the heterozygote will continue to yield all three genotypes. Removal of dominant type individuals is more drastic in its effects because it would include the heterozygotes. There are however cases where dominance is incomplete and the heterozygote appears more nearly intermediate between the dominant and recessive extremes. Under such conditions selection against the dominant phenotype may not extend to some or all of the heterozygotes. If the heterozygotes varied among themselves in their simulation of the dominant phenotype those most closely resembling the recessive might perhaps have the best prospect of survival. And if an organism survives for such a reason some of its progeny could be expected to perpetuate the advantage.

An actual example of such a mechanism was provided by Ford (quoted by Fisher, 1958). A character for wing colour in the currant moth, *Abraxas grossulariata*, follows a basically Mendelian pattern of inheritance but with the heterozygote showing a variable but approximately intermediate expression. By selecting heterozygotes favouring the recessive and raising progeny from them, Ford was able after only three generations to obtain heterozygotes identical in wing colour with the recessives. The incomplete dominance was therefore fairly rapidly eroded. (Selection simultaneously in the opposite direction with related lines of *Abraxas* was able to reinforce the original partial dominance.)

The modification of dominance is reckoned to depend upon many genes of small effect which enhance or reduce the effect of the principal or major gene. An analogy to this might be the significance attached to a word in a sentence which will vary if accompanying words are altered. A particular noun (or major gene) can have its meaning altered, depending on the adjectives (small effect genes or 'modifier' genes) which describe it. The results from *Abraxas* naturally have a significance extending beyond the insects even though in other cases complicating factors can occur.

The individual and its relationships

An individual plant normally results from fusion between two gametes derived from two others. In due time the plant thus produced will itself leave progeny and in so doing will provide only half of any resulting genotype. An exception would be a line of descent maintained exclusively by self-pollination. Each plant is a visible and transitory part of what we may call the 'genetic continuum'. Because it is possible to link the genetics of a particular plant to the genetics of its ancestors and its contemporaries, one can make predictions about its offspring.

An essential part of Mendel's experiments in which he elucidated the principles of segregation and independent assortment was the idea of 'pedigree'. By scrupulous control of pollination Mendel maintained side by side peas with known and separate lines of descent. Under the natural conditions of plant reproduction the idea of pedigree cannot be so readily applied since the lines of descent are not separate but repeatedly and variously interwoven. While the lines of relationship are complex, the properties of genes remain sufficiently unchanged for their behaviour to be forecast statistically. The activities of genes in populations are important and it is therefore necessary now to introduce the Hardy Weinberg Law, a basic proposition of population genetics derived originally for diploid organisms.

Random mating: its maintenance and consequences

The Hardy Weinberg Law may be stated as follows:

If a gene is represented in an infinitely large random mating population by the adaptively neutral alleles A and a in the ratio $qA : (1-q)a$ their proportional frequencies will remain constant.

Since A is present with a frequency of q
and a is present with a frequency of $1-q$

in the cross $Aa \times Aa$

	♂ $\quad qA$	$(1-q)a$
♀		
qA	$q^2\ AA$	$q(1-q)Aa$
$(1-q)a$	$q(1-q)Aa$	$(1-q)^2 aa$

The F_1 progeny consist of

$q^2\ AA + 2q(1-q)Aa + (1-q)^2\ aa$ individuals

the frequency of A at F_1 is now

$q^2 + q(1-q) = q$

and the frequency of a is now

$(1-q)^2 + q(1-q) = 1-q$

The frequencies of A and a thus remain unchanged.

This law makes several assumptions which are:

(i) A and a are of equal selective advantage;

(ii) mating is at random;

(iii) the population is infinitely large;

(iv) migration into or out of the population does not occur;

(v) mutation of A to a or a to A occurs with negligible frequency.

The Hardy Weinberg Law has been influential in evolutionary biology for two reasons. Firstly, it is an elegant and reasonable proposition that relates the fortuitous dispersal of gametes to the apparent constancy of the species. Secondly, if one or more assumptions do not apply in a particular situation there is the prospect of change. It will be at once obvious that each time one of the assumptions does not apply we have what can be a real-life situation. Dominant and recessive genes are seldom of equal advantage, mating may not be at random, some populations are extremely small and clearly mutations must occur with more than negligible frequency. Migration, too, is known to occur.

While population mathematics can be adapted[1] to all manner of novel genetic systems operating among plants (or animals), it should be recognised that the Hardy Weinberg Law is the foundation for much of this work.

The value of population genetics is that it has a predictive function and can in a given situation indicate how the genetic constitution of a population will develop over many generations. Dissociated, however, from adequate field observation, its value is diminished.

Diploidy and evolution

Particularly in the early years of genetics, the subject was almost entirely confined to diploid organisms, and the occasional encounter that the investigator had with oddities like mutations in *Oenothera* and apomixis in *Taraxacum* could only be assimilated after work with diploid organisms had gone sufficiently far.

The diversity and ecological range of diploid plants is remarkable.

---

[1] A comprehensive account of this subject can be found in C C Li's book, *Population Genetics*, University of Chicago Press, 1955.

While one cannot entirely account for this, the foregoing pages indicate something of the flexibility and scope possessed by plants living chiefly in the diplophase. Diploidy permits the coexistence and interaction of various alleles; through meiosis, reassortment is readily achieved and unusual genetic combinations can emerge without jeopardising seriously the future of the entire population.

Chapter 6

Intra- and interspecific boundaries

The classification of plants involves a series of increasingly inclusive groups. Related subspecies together comprise a species and several species constitute a genus. Families and orders are progressively larger groups. The evolutionist concentrates his attention around the species level and it is not difficult to see why this should be. A species is a biological entity which is morphologically and genetically distinct. Above the species level gene exchange or hybridisation between taxa becomes increasingly difficult and within a species occur those genetic mechanisms that may be adaptive or likely to lead to the formation of a new species.

Two separate but related species will normally differ in several respects and the divergence between them can be due to one or more of several causes. To take an obvious example, the tetraploid cotton, *Gossypium hirsutum*, was possibly derived from two diploid species (*G. herbaceum* and *G. raimondii* or their near relatives) by hybridisation and chromosome doubling. The two constituent diploids achieved their own distinctiveness by another route, showing that there are different speciation mechanisms.

Establishment of divergence

Most species are present in a flora as populations. Cross-pollination usually, and self-pollination occasionally, will be for almost all but self-incompatible and dioecious species the pattern of sexual reproduction. This means that an unusual genotype appearing by simultaneous segregation of several homozygous recessive alleles can just as easily disappear. If, therefore, a novel genotype is to survive it could do so more easily if protected from the effects of random recombination. If a diploid crosses with a related allotetraploid the progeny is at once isolated from its parents, since usually only sterile triploids can result. This is isolation without distance. It is genetic.

Apart from the exceptional case of polyploidy it is arguable whether isolation *must* precede the establishment of divergent taxa. What is evident is that isolation would assist divergence and that, in nature, related species seem often to have developed in isolation from each other. It does not follow that isolation *automatically* promotes divergence. *Thespesia populnea* occurs on most tropical coasts in both the New and Old World, where the environment is suitable but, according to Good (1964), with little evidence of segregation.

Isolating mechanisms

Isolating mechanisms may be classified in a variety of ways, of which the following is an example:

Geographical isolation

Plants which are biologically unisolated and would be able freely to exchange genes can be prevented from doing so by spatial separation. If separation is prolonged other isolating mechanisms may accumulate.

Biological isolation

This can operate in many ways either before or after formation of a new sporophyte generation at fertilisation.

PRE-FERTILISATION

(i) Seasonal. Otherwise compatible plants may be mutually isolated by different times of pollen shedding and stigmatic receptivity.

(ii) Ethological. Pollinators confine their activities to one of two populations—thereby preventing pollen transfer between them.

(iii) Incompatibility. Pollen although transferred from one species or subspecies to another will not germinate or will not penetrate the style sufficiently to reach the egg cell.

POST-FERTILISATION

(i) Embryological. The zygote when formed cannot complete its development in the embryo sac.

(ii) Hybrid sterility. An interspecific hybrid if it is produced may develop to maturity and flower but be infertile through failure of pollen or ovules or both.

(iii) Inviability. A hybrid F_1 may be sufficiently fertile to produce F_2 offspring which prove to be weak or inviable.

(iv) Ecological. Two species or subspecies may be able to hybridise successfully but, where the ecological preferences of the parents contrast sharply, absence of intermediate or suitable habitats may prevent survival of the cross.

Isolating mechanisms may operate singly or jointly. Similarly, failure of embryo development can be due to more than one cause.

Hybridisation in the tropics

Natural hybridisation between recognisably distinct species in the tropics is not uncommon. Natural intergeneric hybridisation, although rarer, is reported for example in '*Ruttyruspolia*', found by Meeuse and de Wet (1961/2), being apparently a hybrid of *Ruspolia hypocrateriformis* var. *australis* and *Ruttya ovata* (Acanthaceae). Among the orchids, *Laelia purpurata* × *Cattleya guttata* var. *leopoldii* is known to occur naturally on Santa Catarina island near Brazil (quoted by Lenz and Wimber, 1959). In the Gramineae natural intergeneric crosses include *Zea mays* × *Tripsacum floridanum* (Farquharson, 1957). Intergeneric hybridisation is normally not an important means of gene flow but the example given here from the grasses raises a special problem which will be discussed in detail in chapter 9. In chapter 8 reference will be found to hybridisation in grasses, particularly with regard to *Dichanthium*.

Following from the summary of isolating mechanisms it can be seen that barriers between species are due to several causes. The results of crosses can be instructive as in the following example. *Ipomoea trichocarpa* ($2n = 30$) is a self-incompatible diploid which will occasionally begin seed development if pollinated by the hexaploid *I. batatas* ($2n = 90$), the sweet potato (Wedderburn, 1967). Development begins more slowly than in *I. trichocarpa* intraspecific crosses and is not completed. One reason for the seed failure in the former cross could be the wide disparity in chromosome numbers but this partial embryogenesis together with morphological evidence goes some way towards establishing *I. trichocarpa* as a possible diploid ancestor of sweet potato.

Hybrid sterility has been frequently recorded and one example is that among bananas of *Musa balbisiana* ($2n = 22$) $\times$ *M. textilis* ($2n = 20$), giving a 21-chromosome sterile hybrid that occurs naturally in the Philippines (quoted by Simmonds, 1962). Another example occurs in the Amaranthaceae in crosses involving *Amaranthus dubius* ($2n = 64$) and *A. spinosus* ($2n = 34$). Spontaneous hybrids were observed by Grant (1959) and by Clifford, who claimed that natural hybrids occur near Ibadan, Nigeria, with a frequency of 20 to 30 per cent of the population. As might be expected, hybrids show greater resemblance to *A. dubius* since it provides almost two-thirds of their chromosome content. Because the 'triploid' appears to be quite sterile the prospects for gene transmission from one species to the other are poor, even though partial separation of the sexes combined with wind pollination increases the chances of interspecific hybridisation here.

In Jamaica two quite distinct species of *Croton* (Euphorbiaceae), *C. linearis* and *C. flavens*, occur and occasionally hybridise. Since the former is dioecious and the latter monoecious it is possible by planting a female of the former in isolation with a plant of the latter to obtain automatic natural interspecies hybridisation. Under these conditions all viable seed removed from the female *C. linearis* gives F_1 hybrids intermediate between the parents (see plate 4).

If an F_1 hybrid retains some measure of fertility both F_2 segregation and backcrosses to parental species become possible and the implications of this will now be discussed.

Introgressive hybridisation

Where two species are sufficiently compatible genetically, there is the prospect of fertility being maintained through a series of generations. If two such species are ecologically separated their successful crossing may require intermediate habitats for the progeny to survive. The classic enquiry of this kind involved two irises of the Mississippi delta, *I. fulva*, a plant of shaded wet clay soils, and *I. hexagona* var. *giganticaerulea*, an inhabitant of exposed tidal marsh. The activities of Man created intermediate habitats that permitted the survival of species hybrids. Repeated natural backcrossing of the hybrids to the parents has resulted in the transfer of a small proportion of genes from one species to the other. The implications of this situation, including the broadening of the genetic base of the recipient species, are described in *Introgressive Hybridisation*, an important book by Anderson, published in 1949.

There are few detailed studies of natural introgressive hybridisation among tropical plants, although opportunities exist for them. An interesting exception is that of Fassett and Sauer (1950) working with *Phytolacca*, a genus colonising disturbed ground in Colombia. *P. rivinoides* is a pale purple-flowered plant having loose racemes and generally found growing in lowland sites. *P. rugosa*, having deep purplish-red flowers on a more compact raceme, grows at higher elevations of between 6000 and 9000 feet. Other differences between them may be summarised as follows:

	P. rivinoides	*P. rugosa*
Number of stamens	10–20	8
Number of carpels	10–16	8
Carpels connate	completely	base only
Tepals (perianth parts)	caducous	persistent
Inflorescence length	40 cm	9 cm
Pedicel length	10 mm	5 mm
Tepal length	2 mm	3 mm

P. rivinoides and *P. rugosa* are thus recognisably distinct species occupying contrasted habitats at different altitudes in the eastern Cordillera. Since *Phytolacca* is a plant of disturbed ground, the impact of agriculture would be to create more habitats for it. As weeds, they seem to have migrated to intermediate altitudes and hybridised so that plants examined throughout the altitudinal range show various degrees of genetic mixing. The authors found that whereas the parental character combinations of tepal, style and disc were relatively coherent, the lengths of inflorescence and pedicel and numbers of carpels and stamens were easily modified by introgression.

For a series of nine characters an index was devised where a resemblance to *P. rivinoides* was assigned 2 points, the intermediate condition 1 and to *P. rugosa* 0. Pure *P. rivinoides* would give a score of 18 and pure *P. rugosa* 0. Figure 9 shows the relation between index value and elevation.

Because the two original habitats differ sharply, it is thought unlikely that the two species will merge. Although genes apparently have moved from one species to the other, climatic selection will tend, largely, to preserve each substantially intact at the extremes of the range.

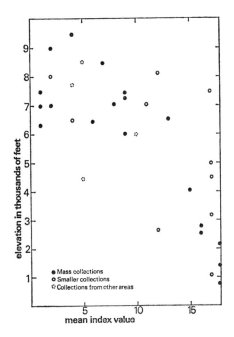

Figure 9 Relation of index value to elevation in *Phytolacca rivinoides–rugosa* series.

Island and mountain floras

Isolation permits divergence and where isolation is most prolonged and intense one might expect striking evidence of change to emerge. High mountains separated by contrasted lowland regions, such as those of East Africa, provide isolated habitats, while other cases are provided by oceanic islands, such as the Galapagos and Juan Fernandez groups. Isolation is not usually just a matter of distance. The mobility of plant propagules can counter the effect of shorter distances.

On the mountains of East Africa are found the curious 'Giant Groundsels' (*Senecio*). Kenya representatives are set out in table 4. The simplest explanation for the origin of these endemic forms is that a common ancestor was formerly more widely distributed. With the advent of warmer conditions it ascended the mountains, became isolated and differentiated into separate species, but it would be interesting to know more about their manner and degree of divergence.

One case worked out in detail is that of the Galapagos tomatoes.

TABLE 4. THE ALTITUDINAL AND GEOGRAPHICAL DISTRIBUTION OF *SENECIO* IN KENYA

Western Kenya Mt Elgon	Cherangani Hills	Central Kenya Aberdare Mountains	Mt Kenya
S. barbatipes (15') 12 500' and above		S. brassiciformis* 10 500–12 500'	S. brassica* 11 500–14 600'
S. elgonensis (25') 10 000–13 200' (and S. barbatipes × elgonensis)	S. dalei* 10 000'	← S. keniodendron (20') → 10 300–13 000' 11 000–14 500' (and S. brassica × keniodendron)	
S. amblyphyllus (25') 9000–10 000'	S. cheranganiensis (30') 8500–10 500'	← S. battiscombei (20') → 11 000–12 500' 11 000–12 500'	

The figure in brackets indicates typical plant height and * indicates dwarfness

Based on Dale and Greenway, 1961

Rick (1963) has provided evidence that the original colonist species, probably from Peru or Equador, was almost certainly *Lycopersicon pimpinellifolium* (a small fruited perennial[1] related to the cultivated tomato) into which had introgressed a few genes from *L. hirsutum*.

Within the Galapagos archipelago the original introduction spread and diversified so that today three distinctive taxa occur : *L. esculentum* var. *minor*, the Galapagos variant of *L. pimpinellifolium* and *L. cheesmanii*, which is in some respects intermediate between the first two. There are other biotypes of more restricted distribution but they seem to be combinations of those just described.

The Galapagos tomatoes are self-compatible and self-pollinating, insect visits being restricted to the infrequent attentions of a bee species, *Xylocopa darwinii*. Such a situation is ideally suited to the establishment of inbred lines and, not surprisingly, seedlings raised in culture from such populations show a high degree of uniformity. Under such conditions, too, the fixation of various alleles can readily occur and has been demonstrated. Several populations at Academy Bay, Indefatigable Island, show complete fixation of j_2, a recessive gene for jointlessness in the pedicel. A population at Wreck Bay, Chatham Island, showed similar fixation for *ag*, an allele governing lack of anthocyanin in the foliage.

The inbreeding habit of *Lycopersicon* suggests at first sight that each island in the archipelago might develop its own distinctive species rather as the composite genus *Scalesia* has done and in a way that recalls the divergence of *Senecio* in East Africa. Among the Galapagos tomatoes this has not happened, partly because of a more efficient inter-island dispersal system. Variation between populations is not much greater between islands than between populations on the same island.

In chapter 5 some of the theoretical properties of diploids were examined and in particular attention was focused on the implications of the Hardy Weinberg Law. In the present chapter something of the field situation has been indicated where the contrast between an original generalisation and the complexities of individual genera should have become apparent. One complicating factor is 'polyploidy' and to this we now turn.

---

[1] Although treated as annuals in cultivation, the edible tomato and its relatives are more nearly perennial in their native habitats, but see chapter 8 pages 73–4.

Chapter 7

Polyploidy

In a diploid organism the reassortment of genes is readily achieved together with the advantages that follow. Linkage reduces the possibilities of recombination, but this serves the long-term interests of the species by preserving intact the greater part of any successful genotype. In both development and reproduction there have been advantages in raising the chromosome number from the haploid to the diploid level for the greater part of the life cycle. Could there be any further advantage from another increase in ploidy?

Among organisms where the diploid sporophyte has achieved predominance this has created no novel problem at meiosis. A polyploid sporophyte, however, overloads the gametogenic cells with homologous chromosomes and disrupts a process which requires *pairs* of homologues. The sexual fertility of a polyploid depends on its ability to simulate the diploid condition and this implies not only orderly chromosome mechanics but profitable recombination too.

It would be inappropriate in this book to devote much space to the general theory of polyploids since this is readily obtainable elsewhere.[1] An attempt will be made instead to examine problems that involve tropical species but have a wider relevance.

[1] An introduction to the cytogenetics of polyploidy is available in *Principles of Genetics* by E W Sinnott, L C Dunn and T Dobzhansky, published by McGraw-Hill (fifth edition 1958). A comprehensive review of plant polyploidy is to be found in chapters 8 and 9 of *Variation and Evolution in Plants* by G L Stebbins, published by Columbia University Press (1950).

Polyploidy and plant habit

Discussions of polyploidy occasionally begin from the frequency with which they occur in a particular flora, but before embarking on this it is important to realise that the extent to which a plant can tolerate polyploidy (and the sexual infertility it can entail) depends upon the means of vegetative propagation available. Stebbins (1938) pointed out that the frequency of polyploidy is highest in herbaceous perennials, intermediate in annuals and least in woody perennials. Thus, where survival through bulbs, corms or tubers is assured and the consequences of disrupted meiosis can be side-stepped, polyploidy may be more easily accommodated.

Polyploidy and plant geography

A situation which recurs in plant geography is one where a family or genus has a fairly circumscribed area of distribution with the exception of perhaps one conspicuously aggressive and widespread subgroup. The Droseraceae are predominantly southern and Australian but the genus *Drosera* is world-wide. The Cruciferae are a chiefly temperate family but one species particularly, *Capsella bursa-pastoris*, is widely dispersed throughout the world.

Good (1964) has provided a list of weedy species which have originated in either temperate or tropical regions and become cosmopolitan. For all but one of these species chromosome numbers are available from Darlington and Wylie (1955), although it is important to realise that chromosome counts from all parts of the various species ranges are not available. Cosmopolitan species of temperate origin are *Capsella bursa-pastoris* ($4x$), *Chenopodium album* ($4x$, $6x$), *Erigeron canadensis* ($2x$), *Euphorbia helioscopa* ($6x$), *Plantago major* ($2x$, $4x$), *Poa annua* ($4x$), *Polygonum aviculare* ($4x$, $6x$), *Solanum nigrum* ($2x$, $4x$, $6x$), *Sonchus oleraceus* ($4x$), *Stellaria media* ($2x$, $4x$), *Taraxacum officinale* ($3x$ and other irregular variants) and *Urtica dioica* ($4x$). Cosmopolitan plants of tropical origin include *Asclepias curassavica* ($2x$), *Cynodon dactylon* ($4x$), *Echinochloa crus-galli* ($4x$), *Gnaphalium luteo-album* ($2x$), *Paspalum distichum* ($4x$) and *Portulaca oleracea* ($6x$). One species in this group, *Amaranthus angustifolia*, has not apparently been cytologically examined.

It is obvious that in these wide-ranging species polyploidy is conspicuous especially among the initially temperate species, thereby suggesting two issues—namely the relative frequency of polyploids in floras of different latitudes and the possibly greater adaptation among polyploids *per se*.

Since the work of Tischler (1935), botanists have been aware of the greater frequency of polyploids at high latitudes. Stebbins (1950) offers three reasons for this. Firstly, the floras of high latitudes have a high percentage of vegetatively reproducing perennial herbs; secondly, such areas have suffered drastic effects of glaciation and therefore hybrids (from which allopolyploids arise), having a relatively wide range of habitat tolerance, would colonise freshly exposed areas in the wake of retreating ice; and thirdly, some polyploids, especially grasses and sedges, may be more resistant to cold conditions than their diploid ancestors. Stebbins (ibid) reported his conclusions about polyploid distribution on the basis of a detailed study involving 100 groups (genera, subgenera or sections of genera). Interestingly, he considered that *no* tropical or Southern Hemisphere groups were well enough known for his purpose.

In view of the foregoing remarks, the work of Morton (1966) is of particular interest. He has shown that for West African plant species the proportion of polyploids is lower than for any other region so far recorded (26 per cent). Individual families, however, may depart appreciably from this general pattern as, for instance, does the Labiatae with 71 per cent polyploid species—a higher figure than is given by the same family in Scandinavia. Morton also reported on the incidence of polyploidy in Cameroons Mountain (49 per cent). These two regions provide a sharp contrast in environmental features, although, of course, both are at tropical latitudes. The West African region generally has experienced long-term stability, whereas Cameroons Mountain has suffered the effects of a more severe climate which has created open habitats. These have been colonised largely by migrant species from the surrounding lowlands together with some of Eurasian affinity.

Plate 5
(a) *Musa acuminata.*
(b) *Musa balbisiana.*
The former inflorescence is not fully pendent. In the first species bracts curl before dropping off, but not in the second.

Plate 6 Illustration showing (a) Tripsacum; (b) teosinte; and (c) maize.
The first two tiller profusely.

Plate 5   (a)   (b)

Plate 6   (a)   (b)   (c)

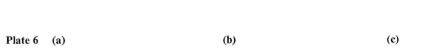

Plate 7

Plate 8

Plate 9

General properties of polyploids

While the effect of increased chromosome number cannot be independent of the particular genes involved, there are several generalisations that can be made about polyploids. These are:

(i) polyploid cells tend to be larger than the corresponding diploid cells and this is reflected in stomatal frequency per unit area, stomatal size and the size of superficial hair cells;

(ii) they often tend to grow more slowly;

(iii) polyploid cells often, but not invariably, show a lower osmotic tension;

(iv) incompatibility systems, both within and between species, sometimes become modified with an increase in ploidy;

(v) autopolyploids generally show a reduction in fertility while allopolyploids may or may not do so. Polyploid pollen grains are sometimes larger than the corresponding diploid ones, but as a consequence of disturbed meiosis there is in the former a higher proportion of abortive grains.

For each of the above examples can be found which are exceptions, but these generalisations about polyploids are based upon work in many species and have wide validity.

Most interest in polyploids centres on the last of the above-mentioned features, namely the various meiotic patterns which occur, thereby reflecting the relationships of the putative ancestors. Stebbins (1947) recognised four major classes of polyploid, which may be outlined as follows:

Plate 7 The reconstructed ancestor of corn, an ear of pod-pop corn (second from right) compared with a modern ear of dent corn (far right) and a prehistoric cob (second from left). The reconstructed ear has female flowers below and male flowers above and in this respect resembles *Tripsacum* (far left). The prehistoric cob also had a male portion which has broken off.

Plate 8 Bisexual inflorescences of Tripsacum and *Zea mexicana* (teosinte) left and centre and the female inflorescence of *Z. mays*. In the two former species, note the terminal male flowers.

Plate 9 An Aztec urn from Mexico (note the maize decoration).

C

Autopolyploids

Where each genome (or haploid set of chromosomes) is derived from the same species or subspecies and the degree of chromosome homology is at a maximum the plant is reckoned to be autopolyploid. The most authentic instances are those produced artificially (by colchicine, for example). Normally meiotic behaviour includes the formation of a large proportion of multivalents.

Segmental allopolyploids

If in a polyploid there are two pairs of genomes that have in common an appreciable amount of homologous chromosome material there is the possibility of a compromise between residual fertility and profitable chromosomal alteration. This group of polyploids is thus partially protected from the consequencies of excessive homology.

Genomic allopolyploids

Where two distantly related species produce a virtually sterile F_1 hybrid, its subsequent elevation to the tetraploid condition ensures that for each distinctive set of chromosomes there is a replica with which it can pair. Fertility is restored, but interchange between dissimilar genomes is minimal.

Autoallopolyploids

These will only occur at the hexaploid level and above, and incorporate some of the features of the previous groups. In a hexaploid, for example, the plant would have four genomes closely alike and two sharply differentiated from these, although similar to each other. Such a plant could arise by hybridisation between an autotetraploid and a not too closely related diploid. Doubling of the resulting triploid would yield a hexaploid of autoallopolyploid type.

Among these four groups the prospects for progressive change are greatest in the segmental allopolyploid group. Residual fertility allows the possibility of perpetuation and chromosomal reorganisation provides alternative genotypes among which natural selection will discriminate. Eventually progeny will result having more regular meiosis, higher fertility and better adapted constitution. Genomic allopolyploids by contrast are relatively more isolated from each parent and backcrossing to either is both less likely and less useful than in the previous case. Moreover, the distinctiveness of the two (or more) genome types prevents much reorganisation within the hybrid. Segregation thus throws few variants and the new taxon perpetuates itself with relatively little change. Examples of these contrasting polyploids are provided by *Solanum tuberosum* (segmental allotetraploid) and *Nicotiana rustica* (genomic allotetraploid). The genus *Saccharum* includes cultivated sugar cane which is cytologically very complex and has features typical of autoallopolyploids.

Endonuclear polyploidy

One other type of polyploidy occurs that does not readily fit into the categories defined by Stebbins. This is 'endonuclear' polyploidy.

Chromosomes in certain representatives of the Cyperaceae and Juncaceae have an unusual type of centromere. Normally the centromere is a discrete point marked by a constriction. Within these families the centromere is diffuse, so that chromosome movement is slightly different. One consequence is that chromosome fragments are capable of directional movement. If, therefore, each of the chromosomes breaks into two separate portions, all of which have the prospect of continued survival, the number of chromosomes is doubled, even though the amount of chromosome material remains constant. Within various genera the result has been to perpetuate an extraordinary range of chromosome numbers. (It is not appropriate automatically to raise each newly discovered chromosome number to specific rank. Where possible it is grouped with the diploid from which it is presumed to have arisen.)

Reducing chromosome numbers

The artificial doubling of chromosome numbers with colchicine is a routine technique among breeders of many plant species but there is as yet no corresponding chemical technique for reduction of chromosome numbers. In a few cases reduction is possible in other ways, but by no means as assured as could be desired. The matter can best be illustrated by a reference to the genus *Solanum* by Hougas, Peloquin and Ross (1958). Ten selections of the common potato ($2n = 48$) were pollinated with diploid selections from *S. phureja*, *S. simplicifolium* and *S. verrucosum*. From 6041 pollinations involving 57 different parental combinations 959 plump seeds were obtained. These eventually yielded 28 haploids ($2n = 24$). The self-incompatibility system whose effectiveness is disrupted in the tetraploid re-emerges in the restored diploid condition.

The example just described is a reduction from tetraploid to diploid. Reduction from diploid to haploid occurs in various genera. A point of terminology is important here. The progeny of a diploid which showed a halved chromosome number would be simply a haploid. Where, however (as in the case of *Solanum* above), a polyploid yields offspring with a halved chromosome number this is referred to as 'polyhaploid'.

Among typically diploid organisms having high fertility and good seed set the occasional haploid occurs and has been reported in the genera *Ipomoea, Cocos, Gossypium, Ricinus, Lycopersicon, Zea, Sorghum* and *Oryza*. In these cases the somatic chromosome number equalled the *assumed* basic number (about which more will be said shortly). These haploids resembled their diploid relatives closely except in being smaller, as a rule, and much less fertile.

The methods by which such haploids arise are varied but not uncommonly accompany 'polyembryony'—the ability of a seed to produce two or more embryos. Polyembryony is attributable to several causes. Simple fission of an embryo would give identical twins of the same ploidy. In other cases fertilisation of the egg cell both gives a diploid embryo and stimulates a synergid to develop as a haploid embryo. The precise nature of the stimulus is not understood.

Haploids are instructive in two respects. The first relates to their role in producing total homozygotes. If a haploid plant is doubled up by (say) a colchicine method each chromosome homologue will be an exact copy of its partner until a mutation eventually occurs. Such a technique would economise the breeder's time, producing pure lines where, for example, a crop had long generation intervals and a large number of valuable recessive genes.

The second interesting feature of haploids concerns their importance in basic number studies. Where a haploid arises from what is assumed to be a diploid, it follows that in the cells of the progeny any chromosome type can theoretically only occur once. If at meiosis only univalents are formed the case is effectively proven. Should the contrary occur and the chromosomes form functional bivalents, then the supposed haploid would be regarded as a 'polyhaploid' and the assumed basic number shown to be a multiple of the true one. Since in the Solanaceae the genus *Petunia* has a basic number of 7 (i.e. approximately 6 for a cytologist exercising his imagination), it could be argued that the basic number of 12 for *Lycopersicon* is really a multiple and that it is derived from a basic number of 6. Subsequently the case would be decided by reference to the ease with which bivalents were formed at meiosis in a tomato with 12 chromosomes per cell and by observing similarities among the chromosomes when examined at pachytene, where the maximum of fine detail is available.

Polyploidy and apomixis

Apomixis is the general term applied to reproductive processes which bypass meiosis and fertilisation. Two major types of apomixis are vegetative reproduction (by corms, bulbs, suckers and so on) and 'agamospermy'. This latter term refers to the asexual formation of embryos and seeds and from casual inspection is not normally distinguishable from ordinary seed development.

As mentioned earlier (p. 63), the existence of vegetative methods of reproduction assists the survival of sterile or nearly sterile polyploids. It appeared, however, at one time that agamospermy is somehow inherently promoted by polyploidy. Certainly the situation where diploids are sexual and related polyploids appreciably agamospermic occurs frequently, but it is an oversimplification to suggest that polyploidy of itself 'causes' apomixis of this kind. The genetic constitution of the original diploids doubtless influences the extent to which sexuality is replaced at higher ploidies.

An interesting, but by no means typical, instance of association between polyploidy and apomixis is provided by an Australian *Casuarina* species, the behaviour of which will now be summarised.

Casuarina—*an unusual breeding system*

The genus *Casuarina* includes about sixty species. About fifteen species form the section Gymnostomae and occur in Malaysia. The remaining Malaysian and all the Australian species make up the larger section Cryptostomae. Only the Australian group has been studied in detail, cytologically, though with interesting results especially in *Casuarina nana*.

The following summary is based on a paper by Barlow (1958).

C. nana, a dioecious species, occurs in New South Wales as isolated populations on exposed ridges. Chromosome number was found to vary within and between populations. Three types of population have been described.

TYPE 1

This population is largely diploid ($2n = 22$) with about equal proportions of males and females. Occasionally aneuploid variants occur but are kept to a low frequency by natural selection.

TYPE 2

Such populations are tetraploid ($2n = 44$) with about equal proportions of males and females. Occasional triploids are not unknown, presumably due to the ingress of fertile pollen from nearby diploids. Tetraploid plants are fertile and reproduction is sexual.

TYPE 3

This is an unusual type of population consisting largely of seed-setting triploid females, but with a small proportion of diploids of both sexes. (Tetraploid females and some highly sterile triploid males occur too but these probably contribute little to the reproductive pattern of the population.)

Seeds of several triploid females from population type 3 were germinated and, in many cases, twin seedlings emerged from *one* seed. Barlow found that in every case of such twins one was tetraploid and the other triploid. The breeding system which makes this possible appears to have the following stages:

(i) the triploid female produces unreduced embryo sacs (meiosis breaks down and the embryo sac has 33 chromosomes per nucleus like the maternal tissue around it);

(ii) fertilisation occurs, involving a haploid male nucleus from a diploid male in the population—a tetraploid zygote is thus formed with 44 chromosomes;

(iii) fertilisation stimulates, in the same embryo sac, a synergid to develop parthenogenetically into a triploid embryo;

(iv) the tetraploid and triploid embryos, embedded in the same mass of endosperm, continue growth, the latter being the more vigorous;

(v) eventually the seed with twin embryos germinates, giving heteroploid seedlings—in time the triploid will outgrow and even eliminate its tetraploid competitor.

The third type of *C. nana* population is complex, since not only does the diploid portion maintain itself sexually but the diploid males are involved in the facultative apomixis of triploids reproducing parthenogenetically. Although the triploids depend upon pollen from diploids, they none the less comprise the largest part of the population.

The significance of polyploidy

The breeding behaviour of *Casuarina nana* is strange, but almost equally odd examples will be mentioned later. Some others might be regarded as disguised polyploids, where high chromosome number accompanies regular meiosis and good seed set and only the existence of diploid relatives discloses their real status. In summary, one can say that while polyploidy can take several forms and provide both highly specialised responses to particular situations and in other cases widely cosmopolitan species it none the less rests on a diploid foundation. Some diploid developments find more effective expression in a polyploid cell environment as, for example, where segmental allopolyploidy leads to greater range and competitiveness.

Small-scale changes of chromosome number

In this chapter no reference has yet been made to alterations in chromosome number involving other than whole genomes. Smaller scale but highly significant alterations in number involving one or two chromosomes are known and a theoretical basis for change of this kind was provided by Darlington (1937). Swanson (1957) contributed a later discussion.

Important and conspicuous in the tropics are the palms. Read (1966) has shown that chromosome numbers in the various palms differ slightly. Examples are: *Chamaedorea alternans* ($n = 16$), *C. microspadix* ($n = 13$), *Gaussia attenuata* ($n = 14$), *Pseudophoenix sargentii* ($n = 17$), *Acrocomia aculeata* ($n = 15$) and *Rhapis excelsa* ($n = 18$). Chromosomes of palms are small, and Read's technique is for this reason noteworthy. He placed palm pollen on colchicine–lactose–gelatine medium which both stimulates germination and holds the chromosomes in a metaphase condition at pollen tube mitosis. In this way counting is done in the haploid stage, thereby simplifying the procedure.

The mechanisms by which the various palms possess different chromosome numbers have not been worked out in detail. Similar changes are known in other families of plants but space does not permit detailed discussion.

Chapter 8

Patterns of evolution in some tropical genera

Variation patterns are to some extent characteristic of a particular genus or family. This is not surprising since a flurry of evolutionary activity may throw up an array of related forms. Among the Myrtaceae there is little variation in basic chromosome number and the incidence of polyploidy is low. In the family Amaryllidaceae the genus *Narcissus* shows a wide range of chromosome numbers, a trend similar to but apparently more extreme than in the less well-known *Zephyranthes*. *Crinum* has a basic number of $n = 11$ and little departure from the diploid condition. Multiple translocation as a way of perpetuating balanced heterozygotes seems at first sight rather hazardous but it functions successfully in *Oenothera* (Onagraceae). Among the Cyperaceae even so fundamental a structure as the centromere is different from that in most other plants and a wide range of chromosome numbers is perpetuated (see chapter 7). The genus *Crepis* (Compositae) is different again. Evolutionary progress seems here to have been accompanied by a *decline* in chromosome number among the diploids and in the Gramineae an immense number of different genera and species have evolved within the confines of a relatively uniform morphology and floral plan. In this chapter, three genera are examined to display something of the operative genetic mechanisms that underlie their variation patterns. One is a dicotyledon (*Lycopersicon*) and two are monocotyledons (*Dichanthium* and *Musa*), the first of these being a grass. Each case has implications for the extent to which a breeder may genetically manipulate representatives of the genus.

Lycopersicon

Within the Solanaceae, two genera, *Solanum* and *Lycopersicon*, show many features in common and an intergeneric hybrid[1] *Lycopersicon esculentum* × *Solanum lycopersicoides* was described by Rick (1951). Notwithstanding this apparent closeness of relationship, the two genera are generally reckoned distinct—the principal point of difference between them being that pollen dehiscence is introrsely by vertical slits in *Lycopersicon* and terminally by pores in *Solanum*. The anthers in the former genus are secured laterally by interlocking hairs in the form of a tube and extend into 'sterile' tips that produce no pollen. In *Solanum* there is a wide range of ploidy (and, associated with this, vegetative reproduction by tubers), whereas in *Lycopersicon* species are virtually all diploid ($2n = 24$). Such polyploids or other chromosome variants as do occur are rare and appear to have contributed nothing to the evolution of the genus.

Since *Lycopersicon esculentum* (the common tomato) is a self-fertile, rapidly growing plant, bearing edible fruits for which there is a ready market, it is not surprising that it has prompted a cytogenetic study perhaps second only to maize in its completeness. Mapping of various genes is well advanced in most of the twelve linkage groups. Varieties of complex parentage and sophisticated breeding are commonplace and the tomato has contributed significantly to the growth of basic genetics.

The genus *Lycopersicon* is of South American origin with representatives in Peru and spreading north towards Ecuador and south to Bolivia and Chile. In this area individual species have a more restricted range. Within the genus a division into two subgroups is made as follows:

Eulycopersicon

Red- or yellow-fruited species, being self-compatible and less obviously perennial in growth habit. Includes the species *L. esculentum*, *L. pimpinellifolium* and *L. cheesmanii*.

Eriopersicon

Green- or white-fruited species, being generally self-incompatible and

[1] These two genera do not normally hybridise. This particular hybrid was obtained by the technique of embryo culture. (After fertilisation but before the seeds are completely developed the embryo from a seed is removed and completes its development on sterile nutrient agar. In this way many difficult hybrid crosses have been raised successfully.)

more obviously perennial in growth habit. Includes the species *L. peruvianum, L. hirsutum* and *L. chilense.*

Because *Lycopersicon* can participate in intergeneric hybridisation it is scarcely surprising that crossing between its subgenera can be accomplished, but only where the female parent is from Eulycopersicon (Rick and Butler, 1956). Within the latter group, hybrids are readily formed, are fertile and between *L. esculentum* and *L. pimpinellifolium* occur naturally (Rick, 1958).

The introduction of the tomato into general cultivation seems to have taken place in the following way. The original centre of diversity was probably the narrow west coast area of the Andes between the equator and latitude 30° S. What is probably a secondary centre of diversity developed on the Vera Cruz–Puebla area of Mexico. Since there is no evidence of pre-Columbian domestication of the tomato in the former centre cultivation probably began in Mexico. If, therefore, *L. esculentum* var. *cerasiforme* spread as a weed from Peru to Mexico this could have provided the initial material. Probably from Mexico after the Conquest, the tomato was taken to Europe some time before 1544 (Jenkins, 1948). Rick (1958) records the interesting development that highly bred American cultivars are now being introduced into the original centre of variability (a situation repeated in other crop plants, notably American maize varieties entering Mexico).

The intensification of tomato culture under modern market gardening conditions has highlighted the susceptibility of tomato to fungus diseases such as early blight (*Alternaria solani*), late blight (*Phytophtlhora infestans*) and grey mould (*Cladosporium fuvum*), to bacterial diseases like canker (*Corynebacterium michiganense*) and wilt (*Pseudomonas solanacearum*), to viruses such as mosaic, spotted wilt and bunchy top, and to insect pests including bollworm (*Heliothis armigera*), thrips (*Thrips tabaci*) and nematodes of the genus *Meloidogyne*. This list by no means exhausts the ills to which the tomato is heir. It does, however, suggest the principal direction of tomato breeding, namely towards increased resistance to diseases and pests, and varieties having multiple resistances increasingly form part of commercial horticulture. The wild species of tomato provide a rich reservoir of genes for various resistances and for the pedigree of a breeder's variety to contain germ plasm from a couple of wild species is not unusual.

As was pointed out earlier, the tomato is diploid and although tomato genetics makes use of various aberrations new varieties are invariably diploid. Not every cultivated species has such a simple genetic system and the two following examples, *Dichanthium* and *Musa*, provide contrasting cases.

Dichanthium

Among tropical grasses belonging to the tribe Andropogoneae is found the 'blue stem' group that is an important ingredient of American pastures. Until recently three taxa were considered sufficiently distinct to merit generic status—*Bothriochloa*, *Capillipedium* and *Dichanthium*. As cytogenetic information accumulated about this group it became apparent that they were more reasonably regarded as one genus under the prior name *Dichanthium*[2] (Harlan—personal note). In a long series of papers by Harlan and his associates the basis of change in *Dichanthium* has begun to emerge.

Within the genus *Dichanthium* it is possible to distinguish four kinds of species (based on Harlan, 1966). These are:

Endemic Indian diploids

A few species appear to be old relics with $2n = 20$. They are confined to very specific stable habitats and seem genetically isolated from each other. Examples include *D. kuntzeanum* and *D. concanensis*.

Australian endemics

There are both diploids and polyploids, sexually reproducing and sometimes cleistogamous. A few forms range beyond Australia to New Guinea and some Pacific islands. An example is *D. sericeum* ($2n = 20$).

American species

High polyploids formerly grouped under *Bothriochloa* occur in the Americas ($2n = 60, 80, 100, 120, 180$ and 220). They are apparently closely interrelated.

Afro-Asian weed species

These are wide-ranging weedy species where $2n = 20, 40, 50, 60$ or 80, and occur in Eurasia, north, east and south Africa, south east Asia and

---

[2] Reference to the original papers mentioned in the text will show in most cases the earlier nomenclature. *Dichanthium intermedium* would appear, for example, as *Bothriochloa intermedia*.

across the South Pacific to Australia. Only the diploids are sexual. The tetraploids are facultative apomicts reproducing sexually at low frequency and the higher polyploids are virtually obligate apomicts. Examples include *D. annulatum* ($2n = 40$), *D. intermedium* ($2n = 40$), *D. ischaemum* var. *ischaemum* ($2n = 40$) and *D. ischaemum* var. *songaricum* ($2n = 60$).

Our concern here is chiefly with the last group.

The species *D. ischaemum* and *D. intermedium* are wide ranging and 'good' taxonomically throughout the greater part of their range (see figure 10). They are aggressive and, typically, colonise disturbed ground. In West Pakistan their ranges overlap and Harlan (1963) reported hybridisation between them. The picture is also complicated because, apart from its hybridisation with *D. ischaemum*, *D. intermedium* in this part of its range also hybridises with *D. annulatum*. From three locations in West Pakistan, Harlan found the following:

Murree Hills

Wide variation was observed including introgression between plants resembling *D. ischaemum* var. *songaricum* and *D. intermedium* and between *D. intermedium* and *D. annulatum*.

Safed Koh

Introgression patterns here resembled those of Murree.

Between Sargodah and Rawalpindi

D. intermedium and *D. annulatum* showed great variability. In addition, a new variant intermediate between *D. annulatum* and *D. pertusum* was found.

The embryo sac of Dichanthium *and hybridisation behaviour*

Hybridisation of the type occurring here is not common although other instances can be found, especially in the Gramineae. It depends partly upon unusual embryo sac behaviour. In this genus there are two kinds of embryo sacs (often in the same ovule). One kind is a normal sexual type, the other acts apomictically. (Figure 11 shows some of the possibilities.)

Figure 10

(a) Sources of collection of *D. ischaemum* maintained at Stillwater, Oklahoma. Circles = *D. ischaemum* var. *ischaemum*; crosses = *D. ischaemum* var. *songaricum*.

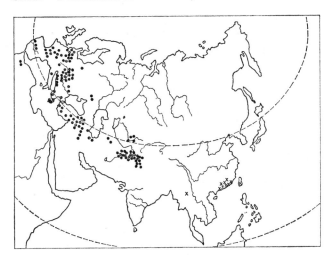

(b) Sources of collection of *D. intermedium* maintained at Stillwater, Oklahoma. Most extensive series of collections are from India and Pakistan, with a sampling from East Africa and Australia. Not all sources could be shown.

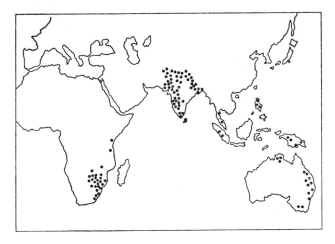

In the case of the hybrid *D. intermedium* × *D. annulatum* which turns out, for example, to be hexaploid, whatever degree of chromo-

Figure 11 Embryo sac behaviour in *Dichanthium intermedium* and the consequences of pollination by *Dichanthium annulatum*. Where fertilisation fails apomictic behaviour follows and the offspring resembles the female parent. If fertilisation succeeds the progeny is intermediate. An exception (the lower case) is where female meiosis fails but is followed by fertilisation. In this case, since the female parent contributes double the male chromosome complement, the progeny resembles the mother more closely.

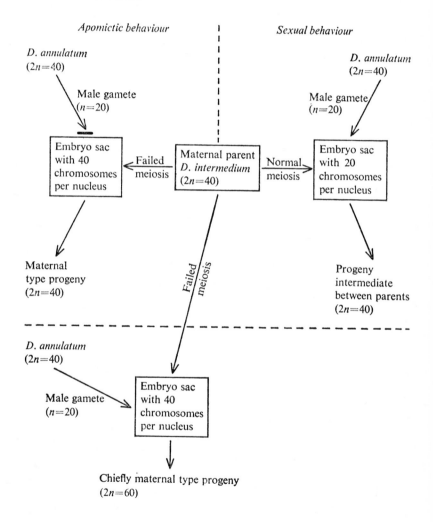

some homology may exist between the different genomes, the plant can, especially as regards vegetative vigour, exploit the attributes of its parents. This somewhat startling introgressive pattern has considerable possibilities. The tendency towards self-sterility combined with the variety of embryo sac performance permits wide crossing and allows the exchange of genetic materials that would probably remain isolated if sexual reproduction alone were operating (Harlan and Celarier, 1961). Indeed a collection of *D. intermedium* assembled from parts of the Old World showed great variability, suggesting wide hybridisation. It is as if one hybridisation made others easier to achieve, allowing *D. intermedium* to 'consume the heredities of its relatives'.

The rather unusual behaviour of *D. intermedium* has led to the idea of a 'compilospecies', the term being used to denote species which are genetically aggressive and capable of assimilating other species (Harlan and de Wet, 1963).

Musa

Musa differs from both *Lycopersicon* and *Dichanthium* in having within one genus a range of basic numbers ($n = 7, 9, 10$ and 11). Banana seeds are hard and gritty and the fruit flesh is opaque. Since, therefore, seediness is highly undesirable, Man has selected for his food forms with few or no seeds. The 'Lacatan' variety for example is completely sterile and parthenocarpic, and thus is highly suitable for cultivation. Another variety, 'Gros Michel', retains a small residual fertility and for this reason can be used in banana breeding, together with various wild species or subspecies that impart disease resistance. The banana breeder is in fact faced with a paradox—namely how to obtain seeds that will give rise to virtually seedless varieties, and reference to the basic genetics of the genus has provided the means to meet this rather daunting requirement. An authoritative and fascinating book by Simmonds[3] has provided much of the material which is included here.

The Musaceae includes two genera, *Musa* and *Ensete*. The genus *Musa* has been divided into four sections, which are as follows:

Eumusa (n = *11*)

Includes the edible bananas and, typically, its representatives have dull bracts, varying in colour from greenish-yellow to dark purplish-brown, and pendent inflorescences.

[3] N W Simmonds, *The Evolution of the Bananas*, Longmans Green (1962).

Rhodochlamys (n = *11*)

This group includes plants which are as a rule smaller than in the previous one and have bright reddish bracts and erect inflorescences.

Australimusa (n = *10*)

Callimusa (n = *10*)

The last two groups have been less fully studied cytogenetically than the eleven chromosome types. There are two other *Musa* species of uncertain taxonomic affinity having basic numbers of 7 and 9 (*M. ingens* and *M. beccarii* respectively). Apart from certain cultivars and experimental clones *Musa* species are normally diploid.

Within *Musa* various crosses have been attempted both between and within sections and among subspecific groups within a species. Intersectional crosses sometimes succeed, such as that between *M. balbisiana* (Eumusa) and *M. coccinea* (Callimusa) and between *M. balbisiana* and *M. textilis* (Australimusa). Both hybrids are sterile and the former is rather weak. The latter occurs naturally in the Philippines. Both these crosses are between $n = 11$ and $n = 10$ types. Among those bananas having a basic number of $n = 11$ inter- and intrasectional crosses have been studied in more detail. The intersectional cross *M. acuminata* (Eumusa) $\times$ *M. laterita* (Rhodochlamys) gives vigorous hybrids whichever way the cross is made. By contrast the intrasectional crosses *M. acuminata* $\times$ *M. basjoo* (within Eumusa) and *M. ornata* $\times$ *M. sanguinea* (within Rhodochlamys) provide in each case scarcely any viable hybrids. Hybrid fertility in crosses between species from Australimusa and Callimusa is much reduced (Shepherd—personal communication). It will be evident that the relationships among sub-generic groups based on morphology and on cytogenetic behaviour are not so straightforward as in *Lycopersicon*.

Of banana species *M. acuminata* and *M. balbisiana* are the most studied. Where these species occur sympatrically, crossing could occur but the degree of reproductive isolation between them is sufficient to prevent introgression. This cross has implications for the evolution of cultivars and will be discussed later. Within *M. acuminata* natural hybrid swarms are known to occur in Malaya, for example, in crosses believed to be between *N. acuminata* sub sp. *microcarpa* and *M. acuminata* sub sp. *malaccensis*.

As regards the origin of cultivated bananas, two species particularly have contributed—in order of importance, *M. acuminata* and *M. balbisiana* (see plate 5). Many cultivars are polyploid and of hybrid origin and if we represent a haploid chromosome set or genome by the initial letter of the species the contributions of each species can be codified. The following examples will suffice.

AA	diploid species *Musa acuminata*
BB	diploid species *M. balbisiana*
AB	edible diploid clone 'Ney Poovan'
AAA	edible triploid clone 'Gros Michel'
AAB	edible triploid clone 'French Plantain'
ABB	edible triploid clone 'Bluggoe'
ABBB	edible tetraploid clone 'Klue Teparod'
AAAA	edible tetraploid clone (a result of banana breeding)

This list does not exhaust the theoretical possibilities but for reasons not entirely understood, *AAAA*, *AAAB* and *AABB* groups, for example, are rare or nonexistent in nature though known experimentally.

The process whereby the cultivated banana evolved has been summarised by Simmonds as follows. The first event was probably the evolution in *M. acuminata* of 'edibility', the combination of parthenocarpy and seed sterility whose joint result is seedless fruit. Parthenocarpy has been shown to involve complementary dominant genes and chromosomal changes such as translocations are also associated with the existence of sterility. Such changes are thought to have occurred in *M. acuminata* sub sp. *malaccensis* and also in other related subspecies. Sterility presumably accumulated gradually and if accompanied by hybridisation among subspecies would have permitted recombination and selection.

The next event to occur was probably the evolution of the *AAA* types and it is known that some nearly sterile edible diploids tend to give triploid progeny. If triploidy added to fruit size and sterility, particularly, Man's selection would favour their survival. Edible bananas of either *AA* or *AAA* constitution were perhaps transported by human agency to places where *M. balbisiana* occurred. Under such conditions several possibilities would have arisen. One is that of hybridisation at the diploid level to give *AB* types. Restitution subsequent to backcrossing could then give both *AAB* and *ABB* individuals.

In more recent times banana evolution has been accelerated under Man's conscious activity and makes use of basic banana cytogenetics. The

D

essentials of the process are to utilise the residual female fertility that exists in the variety 'Gros Michel'[4] and to cross to it AA diploids that have themselves been the subject of a breeding process to improve their disease resistance and other features. Since the embryo sac of the triploid 'Gros Michel' has a characteristic restitution offspring will be tetraploid ($AAAA$). Such material is then tested for agronomic suitability. The principal breeding objectives for banana improvement are increased resistance to Panama disease (*Fusarium oxysporum*), leaf spot or Sigatoka (*Mycosphaerella musicola*), Moko disease (*Pseudomonas solanacearum*) and a burrowing nematode, *Radopholus similis*. At the same time the desirable features of seedlessness, fruit shape and size, yield and storage properties must be retained.[5]

It will be evident that the genetic systems and the resulting options open to the breeder for these three examples are obviously dissimilar. I conclude this chapter, therefore, by drawing attention to some features in common. Firstly, breeding is a complex process that makes full use of whatever basic genetic knowledge is available. Indeed the search for improved crops is a vital stimulus to the progress of fundamental genetics. A further point is that the breeder must take full advantage of material from all parts of the generic range. Either the breeder must travel himself in search of new germ plasm or he must be backed up by plant collectors who are informed about his needs. Lastly, it will be evident that a distinction between wild and cultivated plants is unhelpful if it leads us to exclude from consideration apparently insignificant, unprepossessing or remote species that could add significantly to the vitality of our crops.

---

[4] A variant of 'Gros Michel' named 'Highgate' is of smaller stature and is convenient to use in banana hybridisation. Latterly it has replaced 'Gros Michel' for this purpose.

[5] The development of practical banana breeding in its present phase is conveniently summarised in K Shepherd's paper 'Banana breeding in the West Indies', *PANS* (see bibliography).

Chapter 9

The origin and development of maize

Of all crop plants, maize has been subjected to the most intensive cytogenetic study. Moreover this work has been augmented by information from archaeology, ethnobotany and plant geography. There exists, therefore, an exhaustive literature on its evolutionary relationship to wild grasses, its development at the hands of Amerindian cultivators and the emergence of hybrid corn in United States agriculture.

There is for maize a precise technology that makes it malleable in the hands of plant breeders. The impact of new corn varieties on tropical agriculture—the 'maize homecoming' as it were for the New World—may well prove to be a significant event in twentieth century agriculture. If one adds to substantially improved yields the prospect of repeated cropping during the year production promises to be high. For these reasons therefore the evolution of maize is of particular interest.

Botany

Maize or corn (*Zea mays*) is a panicoid member of the Gramineae belonging to the tribe Maydeae whose species are characterised by unisexual spikelets in either different inflorescences or different parts of the same inflorescence, the male above the female. The tribe contains the following genera, all of them tropical, namely *Coix, Chionachne, Polytoca, Sclerachne, Tripsacum, Trilobachne* and *Zea*.[1] Of these, Coix *aqua-*

---

[1] The two grasses known as 'teosinte', formerly included under *Euchlaena*, are now reckoned to be members of *Zea* (*Z. mexicana*, $2n = 20$ and *Z. perennis*, $2n = 40$).

tica has the lowest chromosome number ($2n = 10$). All the others have a minimum (diploid) chromosome number of $2n = 20$, except for *Polytoca* ($2n = 40$) and *Tripsacum* ($2n = 36$: $2n = 72$ also occurs).

It is reckoned by many but not all workers that maize was unknown in the Old World before the time of Columbus[2] and there seems little reason to doubt that it originated in the New World. More will be said about this later. Two genera of the Maydeae, *Tripsacum* and *Zea*, are indigenous to the New World (see plate 6). Cultivated corn is thought to have derived some of its germ plasm from teosinte and *Tripsacum*.

Present day American representatives of the Maydeae can be distinguished as follows. *Tripsacum* bears its female spikelets below the male spikelets in a terminal spike. Teosinte and corn bear their female spikelets in a unisexual axillary spike (or spikes). In teosinte the spikes are slender, while in corn the spike has a thickened axis called a 'cob'. There are, too, certain associated differences. *Tripsacum* has its inflorescence unprotected by a bract or leaf sheath, whereas in teosinte one protective sheath is present, while in corn there are several. This last character in corn, combined with its non-shattering inflorescence, makes for seed retention rather than dispersal and emphasises its dependence upon mankind for perpetuation. This being so, its evolution can be presumed to post-date Man's entry into the New World, if one accepts the more generally held view of its origin and pre-Columbian distribution.

Man in the new world

No fossils of Man's hominid ancestors have been found in the New World and his presence there is assumed to be due to migration. Of the various routes into the continent the earliest and most likely seems to have been that via the Bering Strait.[3] Man probably first entered the American continent about 20 000 years ago and slowly migrated to the south and east. By about 9500 BC he had reached the tip of South America.

Although in the New World (as in the Old) Man was a hunter, he

---

[2] Scientists are not unanimous on this point and a contrary view has been put by Jeffreys (1964), among others.

[3] There is evidence (and argument) for several other pre-Columbian contacts between the Americas and elsewhere. Some movement between Polynesia and South America seems at least possible, as does contact between Scandinavia and what is now Canada. The entrance via the Bering Strait need not have been the only one but it is almost certainly the earliest.

became under pressure of need a collector of food plants and later a farmer. Willey (1960) supports the view that the change-over occurred first in 'Nuclear America' (the southern two-thirds of Mexico, Andean and coastal Colombia, Ecuador and Peru and adjacent parts of Bolivia). MacNeish (1967) considers that the first signs of incipient agriculture there date from about 7000 BC. Nuclear America is, of course, a vast area and traditions partially isolated are thought to have developed within it. An illustration of this is provided by the first appearance of maize in the agricultural complex of Peru in about 700 BC. This maize was not extremely primitive but rather a well-developed cultivar which it is presumed was introduced from elsewhere (Willey, ibid).

The large significance of maize in the life of Central American peoples is apparent from such things as pottery. Plate 9 shows a vessel on which the maize motif is conspicuous. Not surprisingly, therefore, from botanical and other considerations attempts have been made to trace the origin of this crop in these regions.

Primitive maize

The highly productive commercial corns of United States agriculture are relatively remote from their primogenitors and do not furnish necessarily the best indication of the nature of early maize. Of more value for this purpose are old Mexican races or varieties of corn.

Wellhausen *et al* (1952) have recognised the existence of twenty-five well-defined races of Mexican maize together with four subraces. Of the twenty-five which are believed to have arisen in Mexico four are of particular interest—the so-called Ancient Indigenous Races (Palomero Toluqueno, Arrocillo Amarillo, Chapalote and Nal Tel). These may be described briefly as follows:

Palomero Toluqueno

A short-statured plant producing few tillers. The ear is short, cone shaped, with up to twenty or even more grain rows, the endosperm greyish white. This race is now a high altitude one, occurring in the Mesa Central and the valley of Toluca. It has two distinctive subraces.

Arrocillo Amarillo

This race has very short conical ears, yellow in colour, with about fifteen grain rows. Kernels are of 'pop'[4] type, narrow, thin and awl-

[4] A type of corn where the kernels are small and hard and capable of exploding when exposed to heat.

shaped. Ears occur in twos or threes on each stalk. The race occurs in Puebla and elsewhere between elevations of 1600 and 2000 metres.

Chapalote

Short-statured plant with conspicuous tillering. Although growing best under lowland conditions, it will produce fair-size ears when grown at elevations up to about 1800 metres. The ears are of medium length, about twelve-rowed, rather slender and having small flinty seeds with a chocolate-coloured pericarp. The area of distribution includes the coastal lowlands of Sinaloa and Sonora. It is of interest in that it is both a pop and a weak 'pod'[5] corn.

Nal Tel

This race grows to about two metres tall. It produces few tillers and it is adapted to lowland conditions. The ears are short with about twelve grain rows. The endosperm is flinty, yellow and of the pop type. Its principal location as a pure form is Yucatan.

Other Mexican races include those probably introduced from elsewhere, perhaps from Central or South American sources, those that arose in very early times from hybridisation between these and the Ancient Indigenous types and those that have developed since the Conquest.

Corn is an annual plant whose populations are recreated year by year. To assert that a given race has maintained a particular morphology for several thousand years is at first a surprising claim but such claims rest upon several lines of evidence. Principally they are that:

(i) the so-called Ancient Indigenous Races show upon inbreeding relatively little segregation;

(ii) other races, supposedly more recent, upon inbreeding segregate features recalling the ancient indigenes;

(iii) these particular indigenes are of the pop corn type, which appears from Man's utensils to have been ancient and widespread;

(iv) such prehistoric maize as has been recovered more nearly resembles one or other of the ancient races than it does the more recent ones.

These things do not establish complete certainty but do make the claims seem reasonable.

[5] A type of corn in which the spikelets have long floral bracts (glumes) almost enclosing the kernels.

Since Wellhausen and others published this work more evidence has become available of a most useful kind. Recent studies of prehistoric maize have added considerably to our knowledge and support the above-mentioned work.

Prehistoric maize

The study of maize origins is aided by a most fortunate circumstance, namely that cobs consigned by early Amerindians to trash heaps have in some cases been well preserved in the dry atmosphere of various formerly inhabited caves. Where these refuse deposits have appreciable depth the lowest and highest levels may enclose a considerable time span —sometimes several thousand years. In various cases excavation has shown that the lower (or earlier) levels include cobs noticeably more primitive than those higher up.

The first and most widely known excavations were those made by Dick during 1948 and 1950 at Bat Cave in New Mexico (quoted by Mangelsdorf, 1964). The oldest material recovered from this site was both a pod and pop corn and seems from radio carbon dating to have been growing at about 3600 BC.

Since all manner of genetic variants for maize are available one could, given some kind of hypothesis or model, genetically synthesise a supposed 'primitive' plant. This Mangelsdorf (1958) did by crossing a pod corn with a number of freely tillering varieties of pop corn. There resulted a plant that was both a pod and pop corn with seeds in the tassel, and ears placed at high positions on the stalk which at maturity were not completely enclosed by husks. The way in which these variants would interact was not altogether predictable and so their having been combined provided the archaeologist with a more effective guide in his search for primitive corn (see plate 7).

Since the Bat Cave excavations several others have been made, notably in the Valley of Tehuacan, South of Puebla. Five caves, Coxcatlan, Purron, San Marcos, Tecorral and El Riego, were examined (Mangelsdorf, MacNeish and Galinat, 1964) and found to contain various amounts of food remains. The earliest corn cobs recovered date from 5200 to 3400 BC and are regarded as being those of wild corn, partly because:

(i) they are small and of uniform size;

(ii) the rachises are fragile, as are those of many wild grasses;

(iii) the glumes are relatively long in relation to other structures and must have enclosed the kernels.

The cobs were from 19 to 25 millimetres long whereas modern American Corn Belt hybrids can be more than six times this length. Some of these primitive cobs were apparently once bisexual, having female spikelets below and a stump above, presumed to be where the staminate part of the spike had broken off. Such a feature recalls both *Tripsacum* and the 'primitive' corn synthesised by Mangelsdorf. Further examination of the debris showed that this primitive plant type seems to have persisted until about AD 250. An important observation made earlier in Mexico City was that on a soil core sample maize pollen was found and reckoned to be about 80 000 years old. Since this antedates Man's arrival it could be explained by a maize able to perpetuate itself unaided—a wild maize, in fact.

The presumed wild corn from the Tehuacan caves is not known among present day grasses and has, therefore, almost certainly become extinct. Figure 12 shows the probable appearance of the inflorescence. Mangelsdorf *et al* (1964) consider that there are three important reasons for the demise of wild maize. Firstly, early cultivators took over its habitats and reduced the number and size of wild populations; secondly, pollen from cultivars may have been dispersed and genetically 'swamped' wild corn, imparting to it various features (husked cobs and non-shattering rachises) which would prejudice its survival; and thirdly, the introduction of foraging animals such as goats when the European colonists came would, it is thought, complete the extinction of an already senescent species.

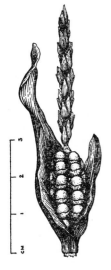

Figure 12 Artist's reconstruction of wild corn based on actual specimens of cobs, husks, a fragment bearing male spikelets and kernels uncovered in the lower levels of San Marcos Cave. The husks probably enclosed the young ears completely but opened up at maturity, permitting dispersal of the seeds. The kernels were round, brown or orange, and partly enclosed by glumes. (Actual size.)

At higher (that is later) levels in the trash heaps were found cob remnants that resembled somewhat those races regarded as ancient indigenes. Cobs, for instance, having affinities with Nal Tel and Chapalote were recovered from the San Marcos Cave and eventually dated AD 200 to 700.

Evolutionary trends in primitive maize

The possible genetic developments that could have occurred in primitive maize were mutation, inter- and intraspecific hybridisation. Apart from other maize varieties, the plants most likely to have contributed germ plasm would be American representatives of the Maydeae—teosinte and *Tripsacum* (see plate 8). Introgression between maize and teosinte can occur in both directions and does so frequently, for example in the maize fields of Mexico and Guatemala. Hybridisation between maize and *Tripsacum* occurs less readily but is hypothesised for some South American varieties of maize. Such a genetic inflow to *Zea mays* would not only create new genotypes but also probably increase its mutability and contribute to its potential for heterosis (Mangelsdorf, 1961). While these things may have gone on spasmodically for thousands of years, the emergence of Man as a cultivator provided a new factor greatly enhancing the survival and spread of genotypes useful to him but otherwise unlikely to perpetuate themselves.

In terms of the habit of *Zea mays* from the dawn of its cultivation until the present, the principal changes have been the increasing separation of the male and female florets (firstly, by their being produced on separate spikes and, secondly, by the ear being produced further down on the plant) and the reduction of tillering. There have, too, been increases in the size and number of the kernels and a replacement of 'pod' and 'pop' characteristics. Husks now enclose the ear and the rachis does not shatter. What maize may eventually become it would be rash to prophesy but already it bears little obvious resemblance to its supposed diminutive and extinct ancestor.

American commercial corn

Modern American corns provide among other things an object lesson in productivity. The pattern of development has been one in which systematic improvement of open-pollinated varieties occurred on a wide scale. This was followed later by the emergence of hybrid corn. The work that Darwin published in 1876 was applied later by an American, W J Beal, on a field scale. Since several comprehensive texts are available on this subject,[6] one can without delay turn to modern tropical corn.

Modern tropical corn

Although maize originated in the tropics, the greatest proportion of its world production today occurs elsewhere. The immense food needs of the tropics will presumably increasingly promote the expansion of corn acreages at low latitudes and stimulate breeders to work in these areas. At present, though, in most tropical countries maize is grown with only minimal skill and, because of the many pests and diseases that attack it, realises only a fraction of its potential yield.

It has become slowly obvious to many people who have pressed modern agriculture upon developing countries that the ingredients for success are subtle, numerous and expensive. It is simply foolish to launch a new variety or crop upon an unprepared peasantry without all manner of efficient supporting services. While, therefore, a new hybrid corn may be available, this in no way guarantees its acceptance. For that part of the world at least, the farmer has usually seen more seasons than the breeder and farming caution is not always grounded in un-thinking habit or timeless superstition. There is, to take a simple illustration, no point in providing an excellent hybrid corn unless its perpetuation from properly maintained pure lines can be assured in-definitely.

The tropical administrator might prefer to leave crop production to a minority of young men with education, aptitude and capital. In practice, he is usually presented with a traditional system of food production that depends on elderly, often illiterate people working, quite diligently, but none the less inefficiently, on small patches of infertile land, who collectively comprise a group that cannot be ignored on either humane or political grounds. If it seems inappropriate to introduce such things here it is not less true that such considerations sharply influence the direction that agricultural development takes in poor countries.

With these things in mind, maize breeders in the tropics cannot always pursue the obvious course of crossing inbred lines to produce single or double cross hybrids, although there is no doubt that this

---

[6] Useful texts include *Corn and Corn Improvement* by G F Sprague, published by Academic Press (1955); *Corn and its Early Fathers* by H A Wallace and W L Brown, Michigan University Press (1956) and *A Professor's Story of Hybrid Corn*, by H K Hayes, Burgess Co. (1963).

would markedly improve maize yields, as an example from Jamaica will show. There the widely grown open-pollinated variety 'J.S.Y.' yields under good management about thirty-five bushels per acre. Under comparable conditions hybrid corn can do much better. A programme by the Pioneer Company, begun in 1963, gave by 1967 locally bred hybrids yielding seventy-three bushels per acre. In similar commercial plantings during 1968 yields of eighty bushels per acre were obtained.

An alternative approach to tropical maize improvement has been used by Wellhausen (1962) and his colleagues working in the Rockefeller Foundation scheme in Mexico. Improved varieties are required for a large number of different habitats and the breeding of individual ones for each ecological area is impracticable. Suppose, however, that two or more distinctive varieties are allowed freely to interbreed to produce what is known technically as a 'composite'. After several generations the result will be a highly diverse and segregating population. If such composites are grown in isolation under carefully defined and well-managed conditions it is possible at each generation to select the most promising and productive individuals and use these to provide seed for the next crop. Over a period of time the result would be, in a given region, the emergence of a local biotype well adapted to that area. This technique of mass selection being simpler to organise than a scheme based on inbred lines and F_1 hybrids holds considerable promise for those areas of the world where breeders are in short supply.

Progress with maize has been impressive, but it is not an isolated instance. Rapid strides have been made too with rice, sorghum, wheat and other crops. Maize need not be the only option open to the farmer at a time when, technically at least, agriculture has the opportunity for expansion in the tropics.

The agro-revolution and the future

The world's food supply provided a continual source of anxiety for about twenty years from the end of the second world war to 1966. A prevailing assumption was that population increase would outrun food supplies. Two interesting results followed. The first was that numerous writers indulged their enthusiasm for anticipating disaster and, secondly (and more usefully), helped to focus manpower and money on an immense complex of technical problems concerned with food production.

Although several countries contributed, the United States has clearly been the principal benefactor. The reasons for this are to be found in her wealth and enterprise and the vast scientific community that are all available. At least as important, however, is the American agricultural tradition which is brisk, progressive and businesslike.

The emphasis on tropical plant breeding has produced several important changes and, since these interact, collectively they are of revolutionary significance. They include:

(i) *Increased fertiliser responsiveness.* The new varieties use fertiliser more efficiently.

(ii) *Aseasonality.* Varieties are now being produced or sought which are less sensitive to seasonal change and permit planting throughout the year.

(iii) *Reduced life cycle.* Newer varieties complete their life cycle more rapidly and allow more generations per year.

(iv) *Mechanisability.* Varieties of crop plants must increasingly be cultivated by machinery thereby reducing labour requirements and allowing both increases in total production and a decline in cost. The newer varieties are ones that can be more easily handled by machinery.

(v) *Increased pest and disease resistance.* Newer varieties are more resistant to various pests and diseases, which both improves performance and economises the use of chemical sprays.

There is today an optimism about the world's food supply which cannot now be written off as frivolity.[7] Perhaps reason rather than reproduction will win. Given a universal system of effective population control, neither need be the loser.

---

[7] For a useful discussion, see L R Brown, *A New Era in World Agriculture* (First annual Senator Frank Carlson Symposium on World Population and Food Supply, Kansas State University, 1968).

Chapter 10

Perspective

A moment's reflection will confirm that the greatest concentrations of plant species and botanists do not coincide. The former occur in the tropics, the latter in the temperate regions. The consequence is that when we examine what information is available about tropical plant evolution we are poorly supplied compared with what can be found elsewhere. It would be no remedy to arrange for botanists to make short forays with limited objectives. What is needed is for more botanists to live and work in the tropics for several years, experiencing at first hand the march of the seasons and observing the intricate interactions of environmental change and plant response. Not every aspect of botany will develop equally in the tropics but one important activity is the continuing appraisal and reappraisal of traditional systematics. To this process a realist sees no end and a taxonomist desires none. Following upon this work comes that of the evolutionist and latterly the plant breeder who recognise the tropics as the area of greatest potential interest.

Tropical plant species

It is evident to biologists that the greatest wealth of species occurs in the equatorial regions, particularly regions that are continually moist, and that both flora and fauna are less abundant and less varied towards the poles. In table 5 two floras are compared for which appreciable data are available. To what extent can such differences be explained?

TABLE 5. COMPARATIVE DATA FOR A TROPICAL FLORA
(JAMAICA) AND A TEMPERATE FLORA (BRITISH ISLES)

Jamaica[1] (*4400 sq. m*)

	Families	Genera	Species
Monocots	34	240	753
Dicots	138	755	2134
	—	—	—
Total	172	995	2887
	—	—	—

British Isles (*about 120 000 sq. m*)

The British Isles flora has been well examined and provides a useful temperate comparison. There are about 125 families, 620 genera and 1900 species, excluding casuals and including native and naturalised species. The precise number of species in the British flora is continually debated. For a useful discussion see Good (1964), chapter 13.

Selected Families

Monocots

Orchidaceae	204 species (61 endemic)	53 species (2–3 endemic)
Cyperaceae	121 species (3 endemic)	110 species (1 ? endemic)[2]
Gramineae	226 species (1 endemic)	146 species (2 ? endemic)

Dicots

Cruciferae	9 species (0 endemic)	91 species (3 endemic)
Euphorbiaceae	112 species (41 endemic)	16 species (0 endemic)
Melastomataceae	78 species (31 endemic)	unrepresented
Rubiaceae	150 species (86 endemic)	19 species (0 endemic)
Compositae	155 species (53 endemic)	183 species[3]

[1] The data by kind permission of Dr C D Adams. (including several endemics)

[2] The genus *Carex* through endonuclear polyploidy is known to add to the number of British species (see p. 67).

[3] One genus, *Hieracium*, through apomixis and specialised segregation adds appreciably to the number of British species and accounts for most of the 'endemics' recorded in the family there. It is arguable, however, whether it is appropriate to use 'endemic' to describe these entities.

There are, of course, certain obvious difficulties in seeking to compare floras quantitatively, although the attempt is instructive. In the British flora, for example, there are reckoned to be some 1500 native species, although divergent views exist about this. The British Isles were largely, though not entirely, glaciated during the Pleistocene and the extent to which species persisted there or were re-introduced across the then existing land bridge with Europe when the ice retreated is still under discussion.

Another difficulty is the climatic variation within each area being compared, and to this can be added the question of the extent to which larger land masses nearby provide migrants to enrich the island floras. It is also true that taxonomists differ as to what constitutes specific rank.

Given all these considerations, the Jamaican flora appears relatively richer than that of the British Isles and reflects a familiar situation—the richer and more varied flora of tropical latitudes when compared with those to the north and south.

During the development of the earth there have been periods of heating and cooling. The former led to an increase in the warmer areas and virtual disappearance of the icecaps. Cooling resulted in an extension of the icecaps towards the equator and a constriction (but not elimination) of what is currently regarded as the distinctively tropical environment. There is, therefore, a contrast between the relatively undisturbed mid-tropical vegetation and the so-called temperate areas that have been subjected to drastic environmental changes. One can thus distinguish between the relative *stability* of tropical areas and those elsewhere that have experienced damaging climatic discontinuity. Richness in tropical species can be partly explained by impoverishment elsewhere but it still leaves unidentified those factors that permit or promote the vast accumulation of species in the tropics.

The wealth of animal species in the continental tropics particularly is partly explained by recalling their ultimate dependence on plant life. The base of the pyramid (photosynthesising organisms) is broader and embraces a more complex array of food chains than elsewhere. This, however, only displaces the problem one remove and emphasises that plant speciation is the key area of enquiry.

Owen (1966), a zoologist, has stressed that tropical species are not simply more numerous, they are, too, more diverse. In plant terms a temperate woodland, for example, consists principally of trees, shrubs and a ground flora whereas a moist tropical forest has these together with an added assemblage of vines, lianes and epiphytes, all of which are represented by several or even many species. Wallace (1891) long ago drew attention to the variety of forms and species coexisting in the tropics and concluded that in the equable equatorial zone there was 'no struggle against climate'. Dobzhansky (1950) has argued that climatic extremes in the temperate regions result in random rather than selective killing and that in the tropics emphasis is upon the individual interorganic struggle.

From the foregoing remarks the nub of the issue—whether equatorial forest more effectively promotes novelty or is simply more tolerant towards it than are other environments—remains largely unresolved. There is, however, another possibility. Bradshaw (1965) considers that phenotypic plasticity is an advantage where environmental changes occur in the lifetime of one individual. The converse of this could be that where conditions of extreme environmental constancy prevail over very long periods plasticity is of little advantage, particularly for a plant of limited distribution. If, therefore, ecological niches were sharply and permanently delineated (as in much tropical rain forest) and within them occurred an intense and prolonged interorganic struggle a wide array of species seems one likely outcome. Provided an organism has adequate reserves of genetic variability, the spectrum of habitats would seem to promote the emergence of novelty. This is, however, speculative.

In chapter 1 an attempt was made to set out the various kinds of environment that occur within tropical latitudes, of which the impressive rain forest forms only one. The work of Morton (chapter 7) has shown that in terms of polyploidy Cameroons Mountain has a different pattern to the nearby lowlands, and to this extent generalisations drawn from tropical rain forest cannot necessarily be extended thus far.

There is a further consideration. Particular families, whether tropical or temperate, and their constituent genera and species are seldom spread uniformly through the environments that will support them. Some families are characteristic of the new world tropics and some of the old. Smaller, more local concentrations of genera and species provide the distribution pattern the botanist normally encounters. While it does not follow that past and present locations are identical, it does seem, for a given genus, that speciation processes are often concentrated in centres of activity.

Concluding comment

In view of the considerable effort that has gone into evolutionary studies, it seems unlikely that many radically new mechanisms for change will emerge from work in the tropics. What does seem likely is that there will be changes in emphasis and in the relative importance of this or that evolutionary process.

Where plant breeders have moved into the improvement of a tropical crop they have seldom resorted to entirely novel methods, although banana breeding could perhaps be an arguable exception. In any event, distinctions between tropical and temperate crops are rather dubious when one considers 'Irish' potato, maize, tomato and cotton, all of which originated in the tropics and have contributed so much to genetics, evolutionary studies and plant breeding. In concluding this book, however, my intention is not to prejudge the various issues but rather to commend them to the reader's curiosity and, better still, his scientific endeavours.

Bibliography

Anderson, E 1949   *Introgressive Hybridisation*
New York, Wiley and Sons. 109 pp

Armour, R P 1959   'Investigations in *Simarouba glauca* Dc. in El
Salvador'
Economic Botany, 13: 41–66

Baker, H G 1955   'Self-incompatibility and establishment after
"long distance" dispersal'
Evolution, 9: 347–8

Baker, H G 1963   'Evolutionary mechanisms in pollination
biology'
Science, 139: 877–83

Baker, H G and
Harris, B J 1957   'The pollination of *Parkia* by bats and its
attendant evolutionary problems'
Evolution, 11: 449–60

Baker, H G and
Hurd, P D 1968   'Intra-floral ecology'
Annual Review of Entomology 13: 385–414

Barlow, B A 1958   'Heteroploid twins and apomixis in *Casuarina
nana* Sieb'
Australian Journal of Botany, 6: 204–19

Bateman, A J 1951   'The taxonomic discrimination of bees'
Heredity, 5: 271–8

Boughey, A S   1957   *The Origin of the African Flora*
Inaugural lecture, College of Rhodesia and
Nyasaland, Oxford University Press. 48 pp

Bradshaw, A D   1965   'Evolutionary significance of phenotypic plasticity in plants'
Advances in Genetics, 13: 115–55

Brian, P W   1959   'Effects of gibberellins on plant growth and
development'
Biological Reviews, 34: 37–84

Brooks, C E P   1949   *Climate through the Ages*
New York, McGraw-Hill. 395 pp

Cope, F W   1962   'The effects of incompatibility and com-
patibility on genotypic proportions of populations of *Theobroma cacao* L.'
Heredity, 17: 183–96

Dale, I R and
Greenway, P J   1961   *Kenya Trees and Shrubs*
London, Hatchards. 654 pp

Darlington, C D   1937   *Recent Advances in Cytology*
Blakiston, Philadelphia, 2nd Ed. 671 pp

Darlington, C D   1958   *Evolution of Genetic Systems*
London, Oliver and Boyd, 2nd Ed. 265 pp

Darlington, C D and
Wylie, A P   1955   *Chromosome Atlas of Flowering Plants*
London, Allen and Unwin. 519 pp

Darwin, C   1839   *The Voyage of the Beagle*
London, Colburn.
The first edition may not be readily available.
An edition more easily accessible is that pub-
lished in 1960 (London, J M Dent. 496 pp)

Darwin, C   1859   *The Origin of Species*
London, Murray.
An edition (1958) has been published with
a foreword by Sir Julian Huxley (New York,
New American Library—Mentor. 479 pp)

Darwin, C   1862   *The Various Contrivances by which Orchids
are Fertilised*
London, Murray. 2nd Ed. 1877. 300 pp

Darwin, C   1876   *The Effects of Cross and Self Fertilisation in the Vegetable Kingdom*
                               London, Murray. 482 pp

Dobzhansky, T   1950   'Evolution in the tropics'
                               American Scientist, 38: 209–21

Dronamraju, K R and   'Constancy to horticultural varieties shown by
Spurway, H   1960   butterflies and its possible evolutionary significance'
                               Journal of the Bombay Natural History Society, 57: 1–8

Du Toit, A L   1957   *Our Wandering Continents*
                               Edinburgh, Oliver and Boyd. 366 pp

Eames, A J   1961   *Morphology of the Angiosperms*
                               New York, McGraw-Hill Co. Inc. 518 pp

Falconer, D S   1960   *Introduction to Quantitative Genetics*
                               Edinburgh, Oliver and Boyd. 365 pp

Farquharson, L T   'Hybridisation of *Tripsacum* and *Zea*'
1957   *Journal of Heredity*, 48: 295–9

Fassett, N C and   'Studies in variation in the weed genus
Sauer, J D   1950   *Phytolacca*. I. Hybridising species in northeastern Colombia'
                               Evolution, 4: 332–9

Fawcett, W and   *Flora of Jamaica*, Vol. V
Rendle, A B   1926   London, British Museum. 453 pp

Fisher, R A   1958   *The Genetical Theory of Natural Selection*
                               New York, Dover Publications Inc. 291 pp

Flory, W S   1944   'Inheritance studies of flower colour in periwinkle'
                               Proc. American Society for Horticultural Science, 44: 525–6

Good, R   1964   *The Geography of Flowering Plants*
                               London, Longmans Green. 3rd Ed. 518 pp

Grant, V   1949   'Pollination systems as isolating mechanisms in flowering plants'
                               Evolution, 3: 82–97

Grant, V   1950   'The pollination of *Calycanthus occidentalis*'
                         American Journal of Botany, 37: 294–7

Grant, V   1958   'The regulation of recombination in plants'
                         Cold Spring Harbour Symposium on Quantitative Biology, 23: 337–63

Grant, W F   1959   'Cytogenetic studies in *Amaranthus*.
                         II. Natural inter-specific hybridisation between *Amaranthus dubius* and *A. spinosus*'
                         Canadian Journal of Botany, 37: 1063–70

Harlan, J R   1963   'Natural introgression between *Bothriochloa ischaemum* and *B. intermedia* in West Pakistan'
                         Botanical Gazette, 124: 294–300

Harlan, J R   1966   'Plant introduction and biosystematics'
                         Plant Breeding Symposium, Iowa State University Press, 55–83

Harlan, J R and   'Apomixis and species formation in the
Celarier, R P   1961   Bothriochloeae Keng'
                         Toronto: *Recent Advances in Botany*. 706–10

Harlan, J R and   'The compilospecies concept'
de Wet, J M J   1963   *Evolution*, 17: 497–501

Hayes, H K   1963   *A Professor's Story of Hybrid Corn*
                         Minneapolis, Burgess Publishing Co. 237 pp

Heslop-Harrison, J   *New Concepts in Flowering Plant Taxonomy*
1953   London, Heinemann Ltd. 135 pp

Hougas, R W,   'Haploids of the common potato'
Peloquin, S J and   *Journal of Heredity*, 49: 103
Ross, R W   1958

Jeffreys, M D W   1964   'Corn in the Old World'
                         Science, 145: 659

Jenkins, J A   1948   'The origin of the cultivated tomato'
                         Economic Botany, 2: 379–92

Jones, D F   1934   'Unisexual maize plants and their bearing on

	sex differentiation in other plants and animals' *Genetics*, 19: 552–67
Lenz, L W and Wimber, D E 1959	'Hybridisation and inheritance in orchids' in C L Withner (Ed.) *The Orchids—a Scientific Survey* New York, Ronald Press Co. 261–314
Li, C C 1955	*Population Genetics* Chicago University Press. 366 pp
Lindauer, M and Kerr, W E 1960	'Communication between the workers of stingless bees' *Bee World*, 41: 29–41, 65–71
MacNeish, R S 1967	'Meso American archaeology' *Biennial Review of Anthropology*, 306–31
Mangelsdorf, P C 1958	'Ancestor of Corn' *Science*, 128: 1313
Mangelsdorf, P C 1961	'Introgression in maize' *Euphytica*, 10: 157–68
Mangelsdorf, P C, MacNeish, R S and Galinat, W C 1964	'Domestication of corn' *Science*, 143: 538–45
Mather, K 1947	'Species crosses in *Antirrhinum*. I. Genetic isolation of the species *majus*, *glutinosum* and *orontium*' *Heredity*, 1: 175–86
Meeuse, A D J and de Wet, J A J 1961/2	'X. "Ruttyruspolia" a natural intergeneric hybrid in Acanthaceae' *Bothalia*, 7: 439–41
Michelbacher, A E, Smith, R E and Hurd, P D 1964	'Pollination of squashes, gourds and pumpkins' *Californian Agriculture*, 18: 2–4
Morton, J K 1966	'The role of polyploidy in the evolution of a tropical flora' in C D Darlington and K R Lewis *Chromosomes Today* Volume One Edinburgh, Oliver and Boyd. 73–76
Muller, H 1883	*The Fertilisation of Flowers* London, Macmillan & Co. 669 pp

Owen, D F 1966	*Animal Ecology in Tropical Africa* Edinburgh, Oliver and Boyd. 120 pp
Petterssen, S 1958	*Introduction to Meteorology* New York, McGraw-Hill. 2nd Ed. 327 pp
Polunin, N 1960	*Introduction to Plant Geography* London, Longmans Green. 640 pp
Pontecorvo, G 1959	*Trends in Genetic Analysis* New York, Columbia University Press. 145 pp
Read, R W 1966	'New chromosome counts in the Palmae' *Principes*, 10: 55–61
Richards, P W 1952	*Tropical Rain Forest; an Ecological Study* Cambridge University Press. 450 pp
Rick, C M 1951	'Hybrids between *Lycopersicon esculentum* Mill. and *Solanum lycopersicoides* Dun' *Proc. National Academy of Sciences* (U.S.), 37: 741–4
Rick, C M 1958	'The role of natural hybridisation in the derivation of the cultivated tomatoes of Western South America' *Economic Botany*, 12: 346–67
Rick, C M 1963	'Biosystematic studies on Galapagos tomatoes' *Occasional Papers, Californian Academy of Science*, 44: 59–77
Rick, C M and Butler, L 1956	'Cytogenetics of the tomato' *Advances in Genetics*, 8: 267–382
Shepherd, K 1968	'Banana breeding in the West Indies' *Pest Articles and News Summaries* Section B, *14*: 37–9
Simmonds, N W 1962	*The Evolution of the Banana* London, Longmans Green. 170 pp
Simpson, G G 1953	*Life of the Past* New Haven, Yale University Press. 198 pp
Sinnott, E W, Dunn, L C and Dobzhansky, T 1958	*Principles of Genetics* New York, McGraw-Hill Co. Inc. 5th Ed. 459 pp

Smithsonian Institution 1951	*Smithsonian Meteorological Tables* Washington. 6th Ed. 527 pp
Sprague, G F 1955	*Corn and Corn Improvement* New York, Academic Press. 283 pp
Stebbins, G L 1938	'Cytological characteristics associated with the different growth habits in the dicotyledons' *American Journal of Botany*, 25: 189–98
Stebbins, G L 1947	'Types of polyploids: their classification and significance' *Advances in Genetics*, 1: 403–27
Stebbins, G L 1950	*Variation and Evolution in Plants* New York, Columbia University Press. 643 pp
Stebbins, G L 1957	'Self fertilisation and population variability in the higher plants' *American Naturalist*, 91: 337–54
Storey, W B 1953	'Genetics of the papaya' *Journal of Heredity*, 44: 70–8
Svedelius, N 1929	'An evaluation of the structural evidences for genetic relationships in plants: Algae' *Proc. International Congress on Plant Science* Ithaca, New York, 1: 457–71
Swanson, C P 1958	*Cytology and Cytogenetics* London, Macmillan. 596 pp
Tai, E A 1958	'Timing the Lacatan banana crop' Banana Board Res. Dept. Jamaica *Occasional Bulletin*, 1: 9 pp
Taylor, B W 1954	'An example of long distance dispersal' *Ecology*, 35: 569–72
Tischler, G 1935	'Die Bedeutung der Polyploidie für die Verbreitung der Angiospermen erlautert an den Arten Schleswig Holsteins, mit Ausblicken auf andere Florengebiete' *Botanische Jahrbuche*, 67: 1–36

Van Steenis, C G G J 1962   'The mountain flora of the Malaysian tropics' *Endeavour*, 21: 183–93

Wallace, A R   1891   *Natural Selection and Tropical Nature; Essays on Descriptive and Theoretical Biology* London, Macmillan and Co. 492 pp

Wallace, H A and Brown, W L   1956   *Corn and its Early Fathers* Michigan State University Press. 134 pp

Wedderburn, M M 1967   'A study of hybridisation involving the sweet potato and related species' *Euphytica*, 16: 69–75

Wellhausen, E J   1962   Plan para el mejoramiento progresivo del maiz en cooperacion con los agricultores. *Sobrietiro del Informe de la Octava Reunión del Proyecto Cooperativo Centro american para el Majoramiento del Maiz*, 10–13

Wellhausen, E J, Roberts, L M and Hernandez, E in collaboration with Mangelsdorf, P C   1952   *Races of Maize in Mexico* Bussey Institution, Harvard University, Cambridge, Mass. 1–223

Willey, G R   1960   'New World pre-history' *Science*, 131: 73–83

Index